MOLECULAR MELODIES

1000 Chemistry Riddles In Verse

UM Digital Creations

Table of Contents

Introduction

Purpose of the Book

Welcome to the fascinating world of chemistry, where the invisible dance of atoms and molecules shapes the very essence of our universe. In this book of riddles, we invite you to explore the wonders of chemical science, unraveling mysteries and discovering the magic that lies within the periodic table and beyond.

Our journey will take us through a landscape of elements, compounds, and reactions. We'll delve into the secrets of acids and bases, peer into the structure of atoms, and witness the transformative power of chemical processes, all in the form of Riddles crafted in a poetic style to make chemistry learning a fun.

As you embark on this chemical adventure, remember that the joy of science lies not just in knowing the s, but in the process of discovery itself. Each riddle you tackle, whether you solve it independently or using the hints provided, is a step forward in your understanding of chemistry.

Don't be discouraged if some riddles seem challenging at first. Chemistry, like any science, rewards persistence and curiosity. As you progress through the book, you'll find your chemical intuition growing stronger, and concepts that once seemed complex will become clear.

How to use the Book

Each page of this book presents riddles, a chemical puzzle designed to challenge your mind and expand your understanding. We've crafted these conundrums to be both entertaining and educational, allowing you to learn while you play.

As you ponder each riddle, you'll find first the "*Hint*" of the riddle .If you find yourself stuck or in need of a gentle clue in the right direction, "Hint" will provide you with a necessary clue to help guide your thinking without giving away the entire solution.Once you've given the riddle your best shot, or if curiosity gets the better of you, look for "*Answer*" at the Answers page of the book.

You will find some of the riddles present in more than one chapter for being relevant at multiple chapters. Don't be overwhelmed by this. It will give you extra practice to learn such topics.

We, UM Digital Creations, has made every effort to avoid errors and omissions in the book. Your suggestions in either pointing out any errors or omissions or for making this book more interactive, will be highly appreciated. Happy reading of Molecular Melodies: 1000 Chemistry Riddles In Verse.

THE BASICS

Chapter 1
Atoms And Elements

1. Tiny world, a universe small,

Nucleus center, electrons crawl.

Protons and neutrons in tight embrace,

What am I, in this microscopic space?

Hint: Think of the building blocks of matter.

2. Lightest of all, I'm number one,

In stars I fuse, that's how they run.

A single proton, that's all I've got,

In water I'm found, believe it or not.

Hint: It's the most abundant element in the universe.

3. Negatively charged, I orbit about,

Without me, ions would surely pout.

I'm shared, transferred, or sometimes free,

In conductors, I move with glee.

Hint: They're fundamental particles with a negative charge.

4. Neutral I stand, in the nucleus I dwell,

My mass like a proton's, can you tell?

I stabilize atoms, keep isotopes unique,

Without charge I remain, neither strong nor weak.

Hint: These particles contribute to atomic mass but not to charge.

5. In the center I sit, positive and proud,

With neutrons beside me, we're quite a crowd.

My number defines the element's name,

In Rutherford's model, I rose to fame.

Hint: These positively charged particles determine the atomic number.

6. I'm unique for each element, a fingerprint of sorts,

My value whole, never fractions or parts.

From hydrogen's one to oganesson's high score,

I determine the element, down to its core.

Hint: This number corresponds to the number of protons in an atom.

7. Round we go, in levels and shells,

Bohr described us, as quantum tells.

Closest to nucleus, we're labeled 'K',

Energy changes as we jump and play.

Hint: These are the regions where electrons are likely to be found.

8. Two's company, eight's a full house,

Noble gases love me, snug as a mouse.

Electrons fill me, then start anew,

Stability's my game, through and through.

Hint: This rule explains why some elements are more stable than others.

9. Protons plus neutrons, together we mass,

Round to the nearest, significant to pass.

I'm noted on tables, right next to symbols,

For isotopes I change, in barely discernibles.

Hint: This whole number approximates the total number of protons and neutrons.Same protons we share, neutrons we do not,

10. Chemically twins, but weights hit different spots.

In nuclear reactions, we're not the same,

Carbon-12, Carbon-14, that's our game.

Hint: These are atoms of the same element with different numbers of neutrons.

11. I'm the key to bonding, a sharing game,

In covalent links, I'm the one to blame.

Lone pairs or bonded, I can be either way,

In Lewis structures, I'm on display.

Hint: These determine how atoms connect in molecules.

12. In pairs we're stable, alone we're free,

In radicals and ions, you'll find me.

I can be shared, or I can roam,

In bonding diagrams, I'm right at home.

Hint: These electrons are not involved in chemical bonding.

13. I'm the force that binds, electrons to core,

The higher the charge, the stronger I soar.

From nucleus to shells, I make my mark,

Ionization energy, I set the bar.

Hint: This electromagnetic attraction keeps electrons orbiting the nucleus.

14. I grow down groups, across periods I shrink,

In atomic radii, I'm the missing link.

Effective nuclear charge, that's my game,

Shielding and attraction, I balance the same.

Hint: This concept explains trends in atomic size across the periodic table.

15. I'm the space where electrons might be,

A probability cloud, not where they'll always be.

Shapes like spheres, dumbbells, and more,

In chemical bonding, I'm at the core.

Hint: These are regions in an atom where electrons are likely to be found.

16. I'm the number that balances, in reactions I shine,

Atoms to molecules, I help align.

Subscript in formulas, I play my part,

In stoichiometry, I'm really smart.

Hint: This number ensures that mass is conserved in chemical equations.

17. I'm the ratio of mass, to volume they say,

For elements and compounds, I vary each day.

In solutions I change, with temperature too,

For identifying substances, I'm a clue.

Hint: This property helps distinguish between different materials, even those in the same state.

18. I'm the path electrons take, around the core,

Circular or elliptical, never a bore.

Bohr proposed me, but quantum improved,

Now probability clouds, have me smoothed.

Hint: In early atomic models, electrons were thought to move in these.

19. I'm the energy gap, between shells so grand,

When electrons jump, photons are at hand.

Absorption or emission, that is the key,

In spectroscopy, it's me you'll see.

Hint: This concept explains the discrete nature of atomic spectra.

20. I'm the change in nucleus, when stability's lost,

Particles emitted, at nature's cost.

Alpha, beta, gamma, I come in three,

Half-life's my measure, of activity.

Hint: This process occurs in unstable atomic nuclei.

21. I'm the model of atoms, like plum pudding they said,

Positive charge spread out, electrons embedded.

Thomson proposed me, before nucleus was known,

In science history, I've grown and grown.

Hint: This early atomic model was disproved by Rutherford's gold foil experiment.

22. I'm the force that repels, when like charges meet,

In atomic nuclei, I bring the heat.

Protons feel me, and try to break free,

But strong nuclear force, won't let them be.

Hint: This force acts between particles with the same electric charge.

23. I'm the blueprint of atoms, a code of sorts,

Electrons arranged, in shells and cohorts.

Noble gas core, plus valence shell,

Element's behavior, I readily tell.

Hint: This notation shows how electrons are distributed in an atom.

24. I'm the energy released, when electrons come near,

Negative ions form, when I appear.

Halogens love me, noble gases do not,

In electron affinity trends, I'm a central plot.

Hint: This energy change occurs when a neutral atom gains an electron.

25. I'm the state of atoms, when charges don't match,

More or fewer electrons, than protons I catch.

Positive or negative, I can swing either way,

In electrolytes and plasmas, I'm here to stay.

Hint: These are atoms or molecules with a net electric charge.

26. I'm the force that holds, nucleons together,

Stronger than Coulomb, I last forever.

But only short-range, I quickly fade,

In nuclear physics, I've got it made.

Hint: This is one of the four fundamental forces of nature.

27. I'm the measurement of mass, for atoms so small,

Carbon-12 is my standard, one-twelfth of it all.

Protons and neutrons, I roughly equate,

In periodic tables, I help calculate.

Hint: This unit is used to express atomic weights.

28. I'm the principle that states, position and momentum,

Can't be known precisely, it's quite a conundrum.

$\Delta x \, \Delta p \geq \hbar/2$, that's how I go,

In quantum mechanics, I steal the show.

Hint: This principle sets a fundamental limit to precision in quantum measurements.

29. I'm the energy needed, to break electron free,

From gaseous atoms, to infinity.

First, second, third, my values ascend,

Periodic trends, on me depend.

Hint: This energy is a measure of how tightly electrons are bound to an atom.

30. I'm the rule that guides, how orbitals fill,

Lowest energy first, that's my bill.

But Hund and Pauli, have their say too,

In electron configuration, I see it through.

Hint: This principle determines the order in which electrons fill atomic orbitals.

31. I'm the force that acts, on charged particles,

Attracting opposites, repelling similarities.

Coulomb's law describes me, with precision,

In atoms and molecules, I make the decision.

Hint: This force is responsible for the attraction between electrons and the nucleus.

32. I'm the odd electron out, unpaired and free,

In free radicals, you'll surely see me.

Reactive and unstable, I seek a mate,

In chemical reactions, I can't wait.

Hint: These electrons make some molecules highly reactive.

33. I'm the nucleus' property, that makes nuclei twist,

Angular momentum quantized, I can't be missed.

Half-integer or integer, I come in two types,

In NMR spectroscopy, I show my stripes.

Hint: This quantum property is important in nuclear magnetic resonance.

34. I'm the energy emitted, when electrons drop down,

From higher to lower, I make photons abound.

Discrete and quantized, my lines are sharp,

In flame tests and fireworks, I play my part.

Hint: This phenomenon produces the characteristic colors of elements when heated.

35. I'm the set of numbers, four in total,

Describing electrons, in their orbital.

n, l, ml, and ms, that's my game,

In atomic structure, I bring fame.

Hint: These numbers completely describe the state of an electron in an atom.

36. I'm the pulling power, of protons in the core,

Minus electron shielding, I'm what's in store.

Across periods I grow, affecting size,

In atomic trends, I'm surely wise.

Hint: This concept explains why atomic radius decreases across a period.

37. I'm the zone of atoms, where electrons reside,

In Bohr's model circular, but quantum says I'm wide.

Principal quantum number, gives me my name,

K, L, M, N, in electrons' game.

Hint: These are the main energy levels in which electrons are found.

38. I'm the subshell shape, spherical and true,

One orbital only, that's my due.

Lowest in energy, in each electron shell,

In atomic structure, I cast my spell.

Hint: This type of orbital is present in every energy level.

39. I'm the subshell set, with lobes like a clover,

Three orbitals I have, when electrons roll over.

px, py, and pz, in space I align,

In bonding and hybridization, I'm quite fine.

Hint: These orbitals are responsible for many covalent bonds.

40. I'm the model that failed, when put to the test,

Alpha particles scattered, not all came to rest.

Plum pudding they called me, but I was all wrong,

In atomic history, I didn't last long.

Hint: This model was disproved by Rutherford's gold foil experiment.

41. I'm the space in atoms, mostly I'm clear,

Between nucleus and electrons, I persevere.

Rutherford discovered me, to everyone's surprise,

Now atomic structure, on me relies.

Hint: This discovery changed our understanding of atomic structure.

42. I'm the family of elements, my outer shells the same,

In periodic table columns, I stake my claim.

Reactivity patterns, I help explain,

From alkali to noble, I maintain.

Hint: These vertical columns in the periodic table share similar properties.

43. I'm the tendency to lose, electrons from my shell,

In alkali metals high, in halogens I dwell low.

Opposite to electronegativity, across periods I fall,

In ionic bonds, I have a ball.

Hint: This property measures how readily an atom gives up electrons.

44. I'm the odd one out, in quantum number's set,

Up or down I go, on this you can bet.

Half-integer values, positive or negative too,

In Pauli's principle, I give a clue.

Hint: This quantum number is crucial in the Pauli Exclusion Principle.

45. I'm the symbol of elements, one or two letters strong,

On periodic tables, that's where I belong.

From H for Hydrogen, to Og for Oganesson,

I represent atoms, from dusk to dawn.

Hint: These abbreviations uniquely identify each element.

46. I'm the unit of charge, named after a physicist bold,

602 x 10^-19 coulombs, if truth be told.

In atoms and ions, I'm the charge they bear,

Protons and electrons, have me to share.

Hint: This is the magnitude of charge carried by a proton or electron.

47. I'm the invisible shield, that blocks outer electrons,

From feeling full force, of nuclear weapons.

Inner electrons provide me, I grow down the group,

In ionization energy trends, I'm quite a scoop.

Hint: This effect reduces the attraction between the nucleus and outer electrons.

48. I'm the concept that says, electrons like to pair,

In orbitals they spin, opposite and fair.

No two alike, in quantum states they dwell,

In electron configuration, I cast my spell.

Hint: This principle is fundamental to understanding electron arrangements.

49. I'm the bond that forms, when metals combine,

Electrons flow freely, in this paradigm.

Conductivity high, malleable too,

In solid structures, I see it through.

Hint: This type of bonding explains many properties of metals.

50. I'm the model of orbitals, with lobes like a rose,

Five shapes I have, as chemistry knows.

In transition metals, I'm often seen,

Complex compounds, are where I gleam.

Hint: These orbitals are important in the chemistry of transition metals.

51. I'm the quantum leap, that electrons make,

From ground to excited, energy I take.

Absorption or emission, that is the key,

In atomic spectra, it's me you see.

Hint: This phenomenon explains the discrete lines in atomic spectra.

I'm the sphere of influence, around atoms I spread,

52. In covalent bonds, I'm clearly read.

Half the distance between nuclei, that's my measure,

In molecular geometry, I'm quite a treasure.

Hint: This concept helps predict how atoms will interact in molecules.

53. I'm the quantum property, that makes particles mirror,

Swap them around, nothing could be clearer.

Fermions or bosons, that's how they're classed,

In particle physics, my influence is vast.

Hint: This property determines how particles behave when interchanged.

54. Noble gases have me, with natural ability.

Eight valence electrons, or full s and p,

In chemical bonding, I'm the key to be free.

Hint: This electron arrangement explains the low reactivity of noble gases.

55. I'm the quantum number, that shapes orbitals' fame,

S, p, d, and f, I give them their name.

From 0 to 3, my values do range,

In electron configuration, I arrange.

Hint: This quantum number determines the shape of atomic orbitals.

56. I'm the quantum state, where energy's least,

For atoms and molecules, I'm quite a feast.

Electrons reside here, when not excited,

In spectroscopy, I'm oft cited.

Hint: This is the lowest energy state of an atom or molecule.

57. I'm the visual tool, for electron probability,

Shapes in 3D, show orbitalability.

Nodes and lobes, I clearly display,

In quantum mechanics, I hold sway.

Hint: These diagrams represent the probability of finding an electron in space.

58. I'm the quantum number that orients in space,

Within a subshell, I find my place.

From -1 to +1, through zero I go,

Magnetic fields split me, don't you know.

Hint: This quantum number describes the orientation of an orbital in space.

59.I'm the property of molecules, with uneven charge,

Electronegativity difference makes me large.

Water's my champion, with bent structure true,

In intermolecular forces, I give a clue.

Hint: This property explains why some molecules are attracted to electric fields.

60. I'm the atomic model, with electrons in shells,

Circular orbits, energy levels as well.

Quantum mechanics proved me incomplete,

But in simple explanations, I'm hard to beat.

Hint: This model introduced the concept of quantized energy levels.

61. I'm the measurement precise, of atom's true mass,

Isotopic abundance, I do surpass.

Carbon-12 as standard, I'm relative you see,

In molecular weight calculations, use me with glee.

Hint: This value considers the masses of all isotopes of an element.

62. I'm the quantum effect, that makes atoms jiggle,

Even at absolute zero, I make particles wiggle.

Heisenberg's principle, gives me my due,

In quantum mechanics, I'm fundamentally true.

Hint: This motion persists even at the lowest possible temperature.

63. I'm the atomic phenomenon, when nuclei align,

In magnetic fields, my signals design.

Proton spins flip, emissions ensue,

In medical imaging, I give doctors a clue.

Hint: This technique is used in a powerful medical imaging method.

64. I'm the quantum property, that's binary by nature,

Up or down, no in-between stature.

Pauli's principle uses me, to guide electron pairs,

In atomic structure, I have important affairs.

Hint: This property of electrons is crucial in the Pauli Exclusion Principle.

65. I'm the atomic spectacle, of waves in phase,

Amplifying each other, in time and space.

In lasers I'm crucial, coherent and bright,

Quantum mechanics explains me right.

Hint: This phenomenon is the basis for laser technology.

66. I'm the quantum concept, of electron pairs trading,

In molecular orbitals, electrons parading.

Bonding and antibonding, that's how I go,

In chemical bonding theory, I steal the show.

Hint: This theory explains bonding in terms of molecular orbitals.

67. I'm the principle quantum, that sets the stage,

Energy levels in atoms, I do gauge.

From one to infinity, my values extend,

On periodic table's periods, you can depend.

Hint: This quantum number determines the main energy level of an electron.

68. I'm the electronic jump, from high to low,

Releasing a photon, in spectral show.

Discrete and quantized, my lines are sharp,

In atomic fingerprints, I play my part.

Hint: This process produces the characteristic spectrum of an element.

69.I'm the quantum rule, that guides electron filling,

Maximum multiplicity, is my billing.

Unpaired electrons first, in degenerate states,

In transition metals, I open electron gates.

Hint: This rule explains why some elements have unpaired electrons.

70. I'm the quantum effect, that links particles tight,

Even when separated, their states unite.

Einstein called me spooky, at a distance acting,

In quantum computing, I'm deeply impacting.

Hint: This phenomenon is key to quantum information theory.

71. I'm the atomic model, that's probability based,

Electrons as clouds, not orbits traced.

Heisenberg's principle, gave me my start,

In modern physics, I play a central part.

Hint: This model replaced the idea of electrons in fixed orbits.

72. I'm the subatomic trio, that nuclei form,Two ups and a down, that's my norm.

Bound by gluons strong, I stand stable and true,

In protons and neutrons, I give mass to you.

Hint: These particles are the components of protons and neutrons.

Chapter 2
Periodic Table

1. I'm the chart of elements, a chemist's delight,

Rows and columns, arranged just right.

From metals to nobles, I show them all,

Mendeleev's legacy, standing tall.

Hint: This organizational tool is fundamental to chemistry.

2. I'm the horizontal row, where periods begin,

My number grows as electrons spin.

From one to seven, I count with pride,

Across the table, I'm your guide.

Hint: These horizontal rows in the periodic table show trends in properties.

3. I'm the vertical column, with similar traits,

My electrons' outer shells seal our fates.

From alkali to noble, I classify with ease,

In periodic trends, I aim to please.

Hint: These vertical columns in the periodic table share similar properties.

4. I'm the leftmost group, with one to give,

Highly reactive, that's how I live.

Soft and silvery, I readily oxidize,

In water I dance, as bubbles rise.

Hint: These elements are the most reactive metals.

5. I'm the rightmost group, noble and stable,

Full outer shells make me unable

To react with ease, in nature I'm rare, In lights and signs, I'm used with flair.

Hint: These elements have very low reactivity due to their electron configuration.

6. I'm the staircase on the table, from Boron to Astatine,

Neither metal nor nonmetal, I'm in between.

Sometimes I conduct, sometimes I insulate,

In semiconductors, I'm truly great.

Hint: These elements have properties intermediate between metals and nonmetals.

7. I'm the group that halves, the noble dream,

One electron shy, I'm not what I seem.

Highly reactive, I form salts with ease,

In table salt and bleach, I aim to please.

Hint: These elements are highly reactive nonmetals.

8. I'm the block of elements, where d-orbitals fill,

Colorful compounds are my special skill.

Variable oxidation states, I often show,

In catalysts and pigments, I'm all the go.

Hint: These elements often form colored compounds and have multiple oxidation states.

9. I'm the trend across periods, as protons increase,

Pulling electrons closer, their orbits decrease.

From left to right, I steadily fall,

In bonding theories, I stand tall.

Hint: This trend affects many properties, including ionization energy.

10. I'm the trend down groups, as shells are added,

Electron shielding makes the pull degraded.

Atoms grow larger, it's easy to see,

In periodic properties, I'm key.

Hint: This trend is opposite to the one across periods.

11. I'm the energy required, to remove an electron,

From a gaseous atom, in its ground-state version.

Across periods I rise, down groups I fall,

In ionization trends, I stand tall.

Hint: This property is crucial in understanding an element's reactivity.

12. I'm the group with two, valence electrons to spare,

Harder than alkali, but still reactive, I swear.

In flames I burn bright, with colors so fine,

In fireworks and flares, I love to shine.

Hint: These elements are named after a type of soil.

13. I'm the series at the bottom, labeled with an L,

My f-orbitals filling, with stories to tell.

From Lanthanum to Lutetium, fifteen in all,

In magnets and lasers, I stand tall.

Hint: These elements are named after a Greek word meaning "to lie hidden."

14. I'm the series below lanthanides, actinides by name,

Radioactive elements are my claim to fame.

From Actinium to Lawrencium, in reactors I'm key,

In nuclear physics, you'll surely see me.

Hint: Many of these elements are artificially created.

15. I'm the metalloid, in group thirteen I dwell,

Third most abundant, in Earth's crust I excel.

In pots and pans, I'm light and strong,

In recycling symbols, I've been all along.

Hint: This element is used extensively in the aerospace industry.

16. I'm the transition metal that life can't do without,

In blood I carry oxygen, without a doubt.

Magnetic and strong, I form alloys with ease,

In Earth's core, I'm the main squeeze.

Hint: This element is responsible for the red color of blood.

17. I'm the group sixteen element, vital for life,

In amino acids, I end chemical strife.

Yellow solid form, or gas with rotten smell,

In volcanoes and hot springs, I dwell.

Hint: This element is named after the Latin word for "brimstone."

18. I'm the lightest metal, in water I dance,

Silvery-white and soft, I take my stance.

In batteries and alloys, I'm light and strong,

In mood stabilizers, I've helped all along.

Hint: This element is named after the Greek word for "stone."

19. I'm the group thirteen element, soft and silvery-white,

In soda cans, I'm lightweight and tight.

In fireworks, I burn bright white,

In aerospace, I'm out of sight.

Hint: This element is named after the Latin word for "alum."

20. I'm the transition metal, in group nine I dwell,

I'm the alkaline earth metal, that starts period four,

In bones and teeth, I'm at the core.

In construction, I'm the key to cement,

In fireworks, my red color is heaven-sent.

Hint: This element is named after the Latin word for "lime."

21. I'm the metalloid, in group fourteen I lie,

Gray and crystalline, semiconductor am I.

In solar cells, I harness the sun,

In computer chips, I'm second to none.

Hint: This element is named after the Latin word for "flint."

22. I'm the lanthanide, with atomic number sixty-three,

Named for Europe, that's the key to me.

In red phosphors, I make TVs glow,

In nuclear reactors, my isotopes show.

Hint: This element is named after a continent.

23. I'm the noble gas, first of my kind,

Lighter than air, in balloons you'll find.

In the sun's core, I'm fused with might,

In MRI machines, I make things right.

Hint: This element is named after the Greek word for "sun."

24. I'm the group fifteen element, essential for life,

In DNA and bones, I end chemical strife.

In fertilizers, I help crops grow,

In matches and fireworks, I steal the show.

Hint: This element is named after the Greek word for "light-bearer."

25. I'm the transition metal, often called quicksilver,

Liquid at room temperature, I'm quite the giver.

In thermometers old, I measured with precision,

Now in some lightbulbs, I aid night vision.

Hint: This element is named after the Roman god of commerce.

26.I'm the alkali metal, that starts period three,

Soft and silvery, in salt I'm the key.

In street lamps, I glow yellow bright,

In biological processes, I help nerves ignite.

Hint: This element is named after the Latin word for "soda."

27. I'm the transition metal, in group ten I lie,

Silvery-white and hard, corrosion I defy.

In coins and jewelry, I'm often seen,

In catalytic converters, I keep exhausts clean.

Hint: This element is named after the German word for "copper nickel."

28. I'm the noble gas, ending period five,

Colorless and odorless, in welding I thrive.

In lightbulbs, I help filaments last,

In windows, I insulate unsurpassed.

Hint: This element is named after the Greek word for "lazy."

29. I'm the lanthanide, with atomic number sixty-two,

Named for a continent, my applications aren't few.

In powerful magnets, I'm strongly aligned,

In lighter flints, my sparks are refined.

Hint: This element is named after a continent.

30. I'm the transition metal, in group five I stand,

Blue-gray and ductile, in aerospace I'm grand.

In surgical implants, I'm biocompatible and strong,

In superconducting magnets, I help research along.

Hint: This element is named after Tantalus in Greek mythology.

31. I'm the transition metal, precious and bright,

Silvery-white and rare, I put up a fight.

In catalytic converters, I clean the air,

In jewelry and electronics, I'm beyond compare.

Hint: This element is named after the Spanish word for "little silver."

32. I'm the halogen, last stable in my group,

Dark violet solid, I'm part of the troop.

In disinfectants, I kill germs with ease,

In contrast media, I help doctors please.

Hint: This element is named after the Greek word for "violet."

33. I'm the alkali metal, third in my group,

Soft and reactive, I'm part of the troop.

In fertilizers, I help crops thrive,

In fireworks, my purple flames come alive.

Hint: This element is named after the English word for "pot ash."

34. I'm the transition metal, in group eleven I shine,

Red-orange and ductile, in wires I'm fine.

In electrical conductors, I'm hard to beat,

In ancient coins, I was quite a treat.

Hint: This element is named after the Latin word for the island of Cyprus.

35. I'm the transition metal, in group five I reside,

Hard and ductile, with high melting point I abide.

In steel alloys, I add strength and shine,

In superconductors, I work just fine.

Hint: This element is named after a Scandinavian goddess.

36. I'm the halogen, second lightest of my kin,

Pale yellow gas, in toothpaste I've been.

Water fluoridation knows me well,

In refrigerants, I used to dwell.

Hint: This element is named after the Latin word for "flow."

37. I'm the noble gas, ending period six,

Colorless and odorless, in lighting I mix.

Used in high-speed photography's flash,

In ion lasers, I make quite a splash.

Hint: This element is named after the Greek word for "hidden."

38. I'm the lanthanide, with atomic number sixty-four,

Named after a Finnish chemist, in magnets I soar.

MRI contrast agents know me well,

In microwave technology, I excel.

Hint: This element is crucial in producing strong permanent magnets.

39. I'm the metalloid, in group fifteen I stand,

Toxic and brittle, in alloys I'm grand.

In semiconductors, I'm paired with gallium,

In fireworks, I create white flares' brillium.

Hint: This element is named after the Greek word for "not alone."

40. I'm the alkali metal, that ends period five,

Soft and silvery, in atomic clocks I thrive.

Discovered by spectroscopy, that's my claim,

In treating bipolar disorder, I found my fame.

Hint: This element is named after the Latin word for "deepest blue."

41. I'm the transition metal, in group seven I dwell,

Hard and brittle, in steel I excel.

Purple in potassium permanganate,

In batteries and welding, I'm first-rate.

Hint: This element is named after the Greek word for "magic."

42. I'm the group thirteen element, soft and light,

In aircraft and rockets, I reduce their weight.

Amphoteric in nature, I react with base and acid,

In packaging and foils, I'm simply magic.

Hint: This element is named after a village in England.

43. I'm the transition metal, in group twelve I stay,

Blue-silver in color, in brass I play.

Galvanization knows me quite well,

In batteries and sunscreen, I cast my spell.

Hint: This element is named after the German word for "sharp."

44. I'm the halogen, heaviest stable of my kind,

Dark purple solid, in disinfectants you'll find.

In photography, I once played a part,

In organic synthesis, I'm off the chart.

Hint: This element is named after the Greek word for "stench."

45. I'm the lanthanide, with atomic number sixty-eight,

Named for a Swedish village, in lasers I'm great.

Ferromagnetic at room temperature, that's rare,

In optical fibers, I amplify with flair.

Hint: This element is used in creating pink glass.

46. I'm the metalloid, in group fourteen I rest,

Gray and brittle, in alloys I'm best.

In car batteries, I'm the key player,

In radiation shields, I'm a top layer.

Hint: This element is named after the Anglo-Saxon word for "lead."

47. I'm the alkaline earth metal, that ends period six,

Soft and silvery, with water I mix.

Radioactive, yet in smoke detectors I dwell,

In nuclear batteries, I serve quite well.

Hint: This element is named after the Latin word for "ray."

48. I'm the transition metal, in group six I shine,

Hard and brittle, in drill bits I'm fine.

Highest melting point of all elements pure,

In filaments for lamps, I long endure.

Hint: This element is named after the Swedish word for "heavy stone."

49. I'm the noble gas, third in my group,

Colorless and odorless, in diving I troop.

Used in excimer lasers, that's my fame,

In window insulation, I play my game.

Hint: This element is named after the Greek word for "new."

50. I'm the group fifteen element, essential for life,

Colorless gas, I cause greenhouse strife.

In the nitrogen cycle, I play a key role,

In fertilizers, I help crops grow whole.

Hint: This element forms about 78% of Earth's atmosphere.

51. I'm the transition metal, precious and rare,

Silvery-white, in catalytic converters I'm there.

Resistant to corrosion, I don't easily tarnish,

In jewelry and electronics, I add my garnish.

Hint: This element is often grouped with platinum.

52. I'm the halogen, lightest of my kin,

Pale yellow gas, in water treatment I've been.

Ozone-depleting, when in certain forms,

In pharmaceuticals, I follow new norms.

Hint: This element is named after the Greek word for "pale green."

53. I'm the group sixteen element, vital for life,

Colorless gas, I end respiratory strife.

In combustion, I play a crucial part,

In the ozone layer, I protect with all my heart.

Hint: This element makes up about 21% of Earth's atmosphere.

54. I'm the transition metal, in group eleven I gleam,

Yellow and shiny, in jewelry I'm supreme.

Resistant to corrosion, I don't easily fade,

In electronics and dentistry, I've got it made.

Hint: This element has been valued by humans for thousands of years.

55. I'm the metalloid, in group fifteen I dwell,

Toxic and lustrous, in semiconductors I excel.

With gallium I pair, in LEDs I shine,

In fiber optic systems, I'm simply divine.

Hint: This element is named after the Greek word for "smoke."

56. I'm the alkali metal, that starts period six,

Soft and silvery, with water I mix.

In atomic clocks, I keep precise time,

In alloys and catalysts, I'm quite sublime.

Hint: This element is named after a ruby-red spectral line.

57. I'm the transition metal, in group seven I stay,

Silver-gray and hard, in alloys I play.

In batteries and electronics, I'm pure,

In water treatment, I make it secure.

Hint: This element is named after a German state.

58. I'm the noble gas, fifth in my group,

Colorless and odorless, in lighting I troop.

Used in ion lasers and plasma displays,

In double-pane windows, I insulate always.

Hint: This element is named after the Greek word for "stranger."

59. I'm the lanthanide, with atomic number seventy-one,

Last of my series, in oil refining I'm not done.

In memory chips, I improve efficiency,

In strengthening steel, I add my proficiency.

Hint: This element is named after an ancient name for Paris.

60. I'm the group fourteen element, in life I'm key,
In organic compounds, I'm all you see.

In diamond I'm hardest, in graphite I'm soft,

In nanotubes and fullerenes, I soar aloft.

Hint: This element forms the basis of organic chemistry.

61. I'm the transition metal, in group nine I reside,

Blue-white and lustrous, in super alloys I abide.

In magnets and turbine blades, I add my might,

In electroplating and catalysts, I shine bright.

Hint: This element is named after a mischievous sprite in German mythology.

62. I'm the alkaline earth metal, that starts period four,

Silvery-white and light, in flares I soar.

In fireworks, I burn bright red,

In photography's early days, I was widespread.

Hint: This element is named after a town in England.

Chapter 3
Chemical Bonds

1. I form when atoms choose to share,

A pair of electrons, not a care.

No loss or gain, just mutual need,

To make their outer shells complete, indeed!

Hint: Not a giveaway, this bond means to share.

2. Tiny travelers, spinning 'round,

Negative charges, tightly bound.

In orbitals they like to hide,

Essential for bonds far and wide.

Hint: They have a negative vibe.

3. Two close friends, forever near,

In a bond, they both appear.

Shared by atoms, they travel in twos,

Guess their name, it's not hard to choose!

Hint: Think of them as a duo in the bond.

4. We stick together, don't you see,

In a covalent bond, we both must be.

Holding tight, we make things stable,

We're part of the molecular fable.

Hint: We're the pair that keeps the atoms glued.

5. We stay alone, we don't join in,

Not part of bonds, where do we fit in?

On one atom, we sit so still,

But our presence can alter shapes at will.

Hint: We aren't shared, but still have power.

6. Together we roam, across both lands,

Bridging atoms, hand in hand.

Not just for one, we belong to two,

In covalent bonds, we hold things true.

Hint: These are the electrons both atoms use.

7. Equal sharing, not one gets more,
We're balanced like never before.

No poles or sides, just symmetry here,

Our strength and bond are perfectly clear.

Hint: No one hogs the electrons in this bond.

8. One side pulls with a little more might,

We're not equal, but it's still alright.

A bit uneven, a dipole we form,

In chemistry, this is quite the norm.

Hint: It's like a tug-of-war between atoms.

9. The space between, not too far,
We measure this across the bar.
From one atom's core to the other's heart,
Our distance plays a crucial part.
Hint: It's the distance between the bonded nuclei.

10. To break this bond, you'll need some power,
Measured in joules or kilojoules per hour.
It tells you how strong this link must be,
More energy, more stability.
Hint: It's the energy to break what atoms share.
11. Between three atoms, we form a line,
Measuring us, you'll need to define.
How wide or narrow our connection goes,
Shapes of molecules, this angle shows.
Hint: It's measured in degrees, between three parts.

12. Head-on overlap, strong and true,
I'm the first bond that'll form for you.
Single and solid, holding tight,

I keep the atoms locked just right.

Hint: I'm the strongest, and the first to form.

13. I come second, when one's not enough,

I add to the bond, make it tough.

Sideways overlap, not as strong,

But in double bonds, I belong.

Hint: I form after sigma in double or triple bonds.

14. Just one pair is all we need,

To join two atoms, guaranteed.

A sigma bond, simple and clear,

The most common bond you'll find here.

Hint: One shared pair, no more, no less.

15. Two pairs we share, to bond so tight,

A sigma, pi—our future's bright.

In oxygen, you'll find us near,

In carbon too, we do appear.

Hint: We have one sigma and one pi bond.

16. Three pairs hold tight, we're tough as steel,

Our bond is strong, that's quite the deal.

Found in nitrogen, sturdy and lean,

We're the strongest bond you've ever seen!

Hint: It's rare and strong, with one sigma and two pi bonds.

17. We live in the outermost shell,

The ones that bond, as you can tell.

In chemistry, we rule the game,

Predicting bonds is our claim to fame.

Hint: Electrons that bond live on the edge!

18. Dots and lines, we like to show,

How electrons in a bond will flow.

Draw us right, and you will see,

The way molecules are meant to be.

Hint: It's a dot-and-line diagram for bonds.

19. Eight is great, we always say,

It's the number we like to display.

In outer shells, it keeps us strong,

With eight electrons, we can't go wrong.

Hint: Atoms are happiest with eight electrons.

20. I pull electrons close to me,

In bonds, I claim them eagerly.

If I'm high, I attract with might,

If I'm low, I don't hold tight.

Hint: It's how hard an atom pulls electrons.

21. I show the way that charge will sway,

From one atom to the other, day by day.

With polar bonds, I point and guide,

The molecule's pull, from side to side.

Hint: I'm a vector showing where the electrons lean.

22. Where two atoms meet, our space combines,

We blend orbitals, crossing lines.

Electrons roam in this new space,

A molecular bond we help to trace.

Hint: It's where two orbitals merge in a molecule.

23. When atoms get close, we start to merge,

Their electron clouds begin to surge.

Where we overlap, bonds are born,

It's how covalent ties are sworn.

Hint: It's when atomic orbitals blend to make bonds.

24. When orbitals mix, they form anew,

sp, sp^2, and sp^3 too.

It changes shapes and bond types well,

In molecules, this tale we tell.

Hint: It's how orbitals blend to create new shapes.

25. One sigma bond is all I need,

Two bonds in a straight line, indeed.

Mixing one s and one p,

This hybrid gives linear geometry.

Hint: I form in molecules like CO_2, think straight lines.

26. Three orbitals merge, just right,

To form a triangle, broad but tight.

I bond in shapes that are flat and wide,

In double bonds, I take great pride.

Hint: Found in molecules with double bonds, like ethene.

27. Four directions, I spread my wings,

In tetrahedral shapes, I bring.

Single bonds are my main role,

Like in methane, I make it whole.

Hint: I give methane its tetrahedral shape.

28. Electrons repel, that's how we guide,

The shape of molecules far and wide.

Our theory predicts the angles and form,

In covalent bonds, we're the norm.

Hint: We predict molecular shapes by electron repulsion.

29. The way we arrange, the form we take,

Angles and shapes, we help make.

From linear to tetrahedral bright,

We define molecules just right.

Hint: It's the 3D shape of molecules based on bonds.

30. Distance from the nucleus to the bond,

I measure how atoms respond.

Half the space that atoms share,

In a covalent bond, I'm always there.

Hint: It's half the distance between nuclei in a bond.

31. To break us apart, you need this much,

Energy's the key, it's the clutch.

Measure of strength, a bond's true might,

More energy needed, the stronger the fight.

Hint: It tells you how tough a bond is to break.

32. We're not ionic, not at all,

We share electrons, that's our call.

Molecules we form, gentle and true,

Between nonmetals, we're the glue.

Hint: We're made of molecules, not ions, and we share electrons.

33. A tiny group, we're bound as one,

With shared electrons, we've just begun.

Water, oxygen, sugar too,

These are the forms I come into.

Hint: I'm the smallest unit of a covalent compound.

34. More than one way to draw me right,

Electrons shift from left to right.

Our real form's a blend of all,

Each structure plays a vital role.

Hint: Multiple Lewis structures, but the real form's a hybrid.

35. I'm not fixed, I move around,

In resonance bonds, I can be found.

Freely shared across the scene,

In metals and ions, I intervene.

Hint: Electrons that don't stick to one place, they spread out!

36. Count the valence, then subtract,

What's shared and lone, that's the fact.

We show which atom holds the stress,

Helping to make molecules less of a guess.

Hint: It shows which atom might carry extra charge in a molecule.

37. One atom donates both its pair,

The other brings none, that's how we share.

A special kind of covalent tie,

We bond without asking why.

Hint: One atom gives both electrons in this special covalent bond.

38. In polar bonds, I make my stand,

Uneven sharing, this was planned.

One side's negative, the other more light,

I create dipoles in bonds so tight.

Hint: I cause tiny poles in polar molecules.

39. I tip the scales in a covalent tie,

One pulls harder, don't ask why.

Electrons drift, they're drawn away,

Creating a charge, where none should stay.

Hint: When one atom pulls electrons harder in a bond, I form.

40. Where do they gather, where do they go?

In a covalent bond, it's good to know.

Regions of charge, thick or thin,

This is where the electrons spin.

Hint: It's where electrons are concentrated in a molecule.

41. I have two poles, a tug of war,

Electrons pulled towards one more.

With charges shifted, I'm not quite fair,

In water, you'll find me there.

Hint: Water is a perfect example of me!

42. I'm balanced, even, with no fight,

Electrons shared equally, left and right.

There's no poles, just calm and peace,

In molecules like methane, this does cease.

Hint: No dipoles here, everything's balanced!

43. How tight we hold, how strong we are,

In chemistry, we're the brightest star.

Break us down, and you will see,

Energy defines our potency.

Hint: It's related to how much energy it takes to break a bond.

44. One atom gives, the other takes,

Forming a bond that never breaks.

When a lone pair makes its move,

A bond we call "dative" we prove.

Hint: It's another name for a coordinate covalent bond.

45. We mirror each other, perfectly so,

In molecules, symmetry will show.

Balanced sides, a structure clean,

Symmetry keeps our shape pristine.

Hint: It's when a molecule's structure is perfectly mirrored.

46. I'm made of two, but they're the same,

From oxygen to nitrogen, that's my game.

Identical atoms, in a pair,

In diatomic form, I'm always there.

Hint: Two atoms of the same element join together in me.

47. I'm two atoms, but not alike,

In covalent bonds, we still unite.

Different elements side by side,

Together as one, we will abide.

Hint: Two different atoms form this kind of molecule.

48. Though we are weak, we still connect,

Between molecules, we can affect.

Holding solids and liquids tight,

In chemistry, we're out of sight.

Hint: Forces that act between molecules, not within.

49. Weak and fleeting, I'm hard to see,

In every molecule, I must be.

Even noble gases feel my touch,

Though I don't pull very much.

Hint: The weakest intermolecular force found in all atoms and molecules.

50. I'm not a bond, though strong I seem,

Between hydrogen and others, I gleam.

Oxygen, nitrogen, and fluorine,

In water and DNA, I intervene.

Hint: I'm a special attraction between hydrogen and highly electronegative atoms.

Chapter 4

States of Matter

1. We're the forms that matter takes,

From ice to steam, whatever it makes.

Solid, liquid, gas are three,

Add plasma, and you'll find me.

Hint: These are the different physical forms in which matter can

exist.

2. I'm rigid and firm, with a definite shape,

My particles locked, no easy escape.

Crystals and rocks are what you'll see,

In the states of matter, what could I be?

Hint: This state has the least kinetic energy among particles.

3. I flow and adapt, to any container,

My volume's fixed, but shape's a disclaimer.

From rivers to oceans, I'm easy to pour,

Which state of matter am I, can you be sure?

Hint: This state can flow and take the shape of its container.

4. I expand to fill, whatever space I'm in,

My particles zoom, with vigorous spin.

Invisible often, but winds reveal me,

In states of matter, can you feel me?

Hint: This state has the most kinetic energy among the common three states.

5. I'm the fourth state, with ions to spare,

In stars and lightning, I give quite a flare.

Electrons freed from atomic embrace,

In matter's states, I have my place.

Hint: This state consists of ionized particles and is found in stars.

6. When heat is added, I start to melt,

Solid to liquid, a change that's felt.

Ice cream in summer knows me well,

Which process am I, can you tell?

Hint: This process involves a solid changing to a liquid.

7. I'm the process where liquids turn to gas,

At the surface it happens, with molecules that pass.

Slower than boiling, I cool things down,

In sweat and puddles, I can be found.

Hint: This process occurs at the liquid's surface at any temperature.

8. I'm the point where solid, liquid, and gas,

Coexist together, in equilibrium, they pass.

Water knows me at 01°C precisely,

What am I called? concisely.

Hint: This point represents where all three common states of matter coexist.

9. I'm the change direct, from solid to gas,

No liquid between, as molecules pass.

Dry ice does this, at room temperature,

What's my name? I'm no caricature.

Hint: This process involves a direct transition from solid to gas.

10. I'm the energy needed, to change the phase,

From solid to liquid, at melting's stage.

Ice to water, I'm the heat you seek,

What kind of heat am I? Take a peek.

Hint: This is the energy required to change a substance from solid to liquid.

11. I'm the state between, not liquid, not gas,

Critical temperature and pressure surpass.

Distinct phases gone, I'm a fluid super,

What state am I? I'm quite the trooper.

Hint: This state occurs beyond the critical point where distinct liquid and gas phases cease to exist.

12. I'm the process where gas turns to solid,

Skipping the liquid, going straight to stolid.

Frost on cold windows, I'm the cause,

What process am I? Give me applause.

Hint: This is the reverse process of sublimation.

13. I'm the point where vapor pressure equals,

The atmospheric pressure, boiling fuels.

Change with altitude, I surely do,

What point am I? Give me my due.

Hint: This point changes with atmospheric pressure and affects boiling.

14. I'm the temperature where solid turns to liquid,

For pure substances, I'm fixed and vivid.

0°C for water, 1064°C for gold so pure,

What point am I? Of this be sure.

Hint: This is the temperature at which a solid changes to a liquid.

15. I'm the energy released, when gas condenses,

Or when liquid freezes, as cold commences.

The reverse of latent heat, I'm what you get,

What kind of heat am I? Don't forget.

Hint: This is the energy released when a substance changes to a more condensed state.

16. I'm the process where particles spread,

From high concentration, to low instead.

In gases and liquids, I make things mix,

What process am I? No need for tricks.

Hint: This process involves the movement of particles from areas

of high concentration to low concentration.

17. I'm the force between molecules, a gentle attraction,

In liquids and solids, I cause cohesion's action.

Van der Waals knew me, I'm weak but important,

What force am I? I'm quite the assortant.

Hint: These are weak attractive forces between molecules.

18. I'm the property of fluids, to resist the flow,

Honey has more of me, than water, you know.

Internal friction, is what I bring,

What property am I? I'm no small thing.

Hint: This property measures a fluid's resistance to flow.

19. I'm the tendency of liquids, to minimize their space,

Forming droplets and bubbles, with curved surface grace.

Water beading on wax, shows me clear,

What property am I? I'm quite dear.

Hint: This property causes liquids to form spherical droplets.

20. I'm the pressure exerted, by a confined gas,

On container walls, as molecules pass.

Boyle and Charles knew me, in laws they made,

What pressure am I? I'm not afraid.

Hint: This pressure is caused by gas molecules colliding with container walls.

21. I'm the state of matter, with properties between,

Not quite a solid, but no liquid sheen.

Toothpaste and gels, are examples of me,

What state am I? Can you foresee?

Hint: This state has properties of both solids and liquids.

22. I'm the process where solids change shape,

Under stress they bend, no need to gape.

Elasticity and plasticity, are types of me,

What process am I? Let's agree.

Hint: This process involves changes in solid shape due to applied forces.

23. I'm the bouncing back, of solids under stress,

Remove the force, and I'll do the rest.

Springs and rubber bands, know me well,

What property am I? Can you tell?

Hint: This property allows materials to return to their original shape after deformation.

24. I'm the permanent change, in a solid's form,

Once deformed, there's no going back to norm.

Modeling clay knows me, as do metals bent,

What property am I? To this, assent.

Hint: This property involves permanent deformation of a material.

25. I'm the random motion, of particles suspended,

In fluids they dance, their paths extended.

Pollen in water, shows me clearly,

What motion am I? Name me sincerely.

Hint: This motion is named after the botanist who first observed it in pollen grains.

26. I'm the transition smooth, no clear boundary line,

Between states of matter, I blur the design.

Critical point passed, distinctions fade,

What kind of transition am I? Don't be afraid.

Hint: This transition occurs without a clear distinction between phases.

27. I'm the property of solids, to resist a scratch,

On Mohs scale I'm measured, a mineral's match.

Diamond's at ten, talc at one, so soft,

What property am I? I'm discussed oft.

Hint: This property measures a material's resistance to scratching.

28. I'm the clumping together, of particles fine,

In colloids and suspensions, I form design.

Milk curdling shows me, as does blood clotting,

What process am I? I'm worth spotting.

Hint: This process involves particles in a colloid coming together.

29. I'm the ability of solids, to be drawn to wire,

Metals have me most, as smiths require.

Gold's the best at me, copper's not bad,

What property am I? Metalworkers are glad.

Hint: This property allows a material to be drawn into a wire.

30. I'm the layer formed, where liquid meets air,

Surface tension's my game, I'm quite fair.

Insects can walk on me, a feat quite neat,

What am I called? I'm no small feat.

Hint: This layer has special properties due to the interface between liquid and air.

31. I'm the process where liquids climb,

Against gravity, taking their time.

In plant stems and paper towels, I'm seen,

What process am I? I'm quite keen.

Hint: This process involves liquids moving up narrow spaces without external forces.

32. I'm the unusual solid, with no crystal to show,

Disordered structure, my particles don't grow.

In glass and plastics, I'm commonly found,

No sharp melting point, I soften around.

Hint: This type of solid lacks a regular crystal structure.

33. I'm the process that turns, a gas to a solid,

In air fresheners, I keep scents valid.

Carbon dioxide knows me, as dry ice I form,

In cloud seeding, I help weather transform.

Hint: This process involves cooling a gas directly into a solid.

34. I'm the liquid that flows, with no viscosity,

Quantum state of matter, full of curiosity.

At temperatures near zero, helium's my game,

In studying quantum effects, I've found my fame.

Hint: This unique state of matter has zero viscosity and infinite thermal conductivity.

35. I'm the state of matter, electrons tightly bound,

In neutron stars, I can be found.

Densest known matter, in the universe so far,

In astrophysics, I'm quite the star.

Hint: This exotic state of matter is found in extremely dense stellar remnants.

36. I'm the process that happens, when gases cool down,
On grass in the morning, as dew I'm found.
Water vapor to liquid, I make the shift,
In weather and distillation, I give nature a lift.
Hint: This process involves cooling a gas to form a liquid.

37. I'm the point where solid, turns straight to vapor,
No liquid in between, I'm quite the caper.
For iodine and camphor, I'm easily seen,
In chemical purification, I keep things clean.
Hint: This is the temperature at which a solid directly becomes a gas.

38. I'm the liquid crystal, neither fully solid nor fluid,
In calculators and watches, I'm far from stupid.
Align in electric fields, changing what you see,
In mood rings and TVs, I'm key to the spree.
Hint: This state has properties of both liquids and crystals.

39. I'm the process where liquid, becomes more ordered,
Forming a solid, as temperature's lowered.
In ice cube trays, I'm easy to do,
In metallurgy, I give strength anew.

Hint: This process involves cooling a liquid to form a solid.

40. I'm the unusual matter, where atoms play cool,

Near absolute zero, I break every rule.

Bose-Einstein's my name, a condensate so rare,

In quantum physics, I'm beyond compare.

Hint: This exotic state of matter occurs at extremely low temperatures.

41. I'm the process that happens, in your refreshing drink,

When gas leaves solution, making bubbles, I think.

In carbonated beverages, I cause the fizz,

In deep-sea diving safety, I'm serious biz.

Hint: This process involves gas coming out of a liquid solution.

42. I'm the cloudy mixture, not solution, not pure,

Particles suspended, visibility obscure.

In milk and fog, I'm easy to see,

In industrial processes, I'm key to many.

Hint: This mixture has particles that are larger than those in a solution.

43. Shaping materials, I'm par for the course.

Hint: This process involves heating a solid until it becomes pliable but not liquid.

44. I'm the state of matter, with properties most odd,

Neither gas, liquid, solid, I leave scientists awed.

Fermionic particles, in quantum degeneracy,

In white dwarfs, I exist with high energy.

Hint: This state occurs when matter is compressed to extreme densities.

45. I'm the process that slows, molecular motion,

As temperature drops, I cause commotion.

In refrigerators, I'm the cooling trick,

In cryogenics, I work quick.

Hint: This process involves reducing the kinetic energy of particles.

46. I'm the change that occurs, when crystals realign,

No state change needed, just pressure fine.

In geology, I create metamorphic rock,

In metallurgy, I strengthen the stock.

Hint: This process changes a solid's crystal structure without melting.

47. I'm the process where gas, gets squeezed tight,

Volume decreasing, pressure at height.

In bike pumps, I'm put to the test,

In refrigeration cycles, I never rest.

Hint: This process involves reducing the volume of a gas, increasing its pressure.

48. I'm the mixture of two states, liquid and gas,
In boiling water, I come to pass.
Bubbles rise up, through the liquid phase,
In distillation columns, I set the pace.
Hint: This occurs when a liquid is at its boiling point.

49. I'm the process that happens, when droplets combine,
Growing ever larger, as they align.
In rain formation, I play a part,
In chemical separations, I'm rather smart.
Hint: This process involves small liquid droplets joining to form larger ones.

50. I'm the state of electrons, in metals so free,
Flowing with ease, conductivity's key.
Not bound to atoms, I roam around,
In electrical circuits, I'm gladly found.
Hint: This describes how electrons behave in metallic bonding.

51. I'm the process where solids, get smaller in size,
No state change needed, just time as allies.
In pharmaceuticals, I increase dissolution rate,
In cooking, I help flavors integrate.

Hint: This process reduces the particle size of a solid.

52. I'm the point of balance, where changes reverse,

Melting or freezing, could occur or disperse.

For water, I'm zero, on Celsius scale,

In phase diagrams, I'm a crucial tale.

Hint: At this temperature, a substance can exist as both solid and liquid.

INORGANIC

CHEMISTRY

Chapter 5

Acids and Bases

1. I make water swell with ions galore?

H+ for acids, OH– and more.

A Swedish chemist, I made the rule,

To classify compounds, simple but cool.

Hint: I'm the foundation of classic acid-base views.

2. I donate or accept, depending on the case?

Proton exchange, in the chemical race.

Acids give, while bases receive,

A simple idea, can you believe?

Hint: Think of who's giving or taking the H+.

3. I don't deal with protons, not in my lane?

Electrons are my game, simple and plain.

An electron pair I happily share,

Or I accept one, if you're fair.

Hint: Forget the protons, focus on electron pairs.

4. I measure how strong or weak you can be?

For acids and bases, just look to me.

From zero to fourteen, a scale you can trust,

Hint: A balance of H+ and OH– defines me.

5. When an acid loses, it gains a new friend?

A base takes its place, a shift in the end.

Like dance partners swapping, we move in pairs,

Hint: Every acid has a partner after giving up an H+.

6. I give my H+ with the greatest of ease?

In water I dissociate, as quick as a breeze.

My strength is unrivaled, that much is clear,

Hint: No hesitation in letting go of protons.

7. I hold onto my protons with some degree?

Not all are released, only partially.

In water I linger, not fully undone,

Hint: I'm not eager to dissociate completely.

8. I release OH– without delay?

Dissolving in water, I clear the way.

I neutralize acids without a fight,

Hint: OH– flows freely when I'm around.

9. I hover and linger, not too fast?

I don't grab protons or release OH– at last.

A weak reaction, a slower pace,

Hint: I don't rush to accept H+ or release OH–.

10. Drop by drop, I measure with care?

To find the point where two meet fair.

An indicator tells me the story,

Reaching equivalence, I find the glory.

Hint: I help find the perfect balance in chemistry.

11. I keep things steady, no drastic change?

With acids or bases, I stay in range.

I'm resistant to swings, stable and strong,

Hint: I keep pH in check when things get acidic or basic.

12. I mix with water, and acids I create?

My oxide form isn't up for debate.

Hint: I react with water to make a sour solution.

13. When I meet water, a base is born?

OH– is formed, I've always sworn.

Hint: Think of me in the company of alkaline reactions.

14. I'm a little confused, I play both sides?

In acids and bases, I can abide.

Hint: I'm neutral, but can be swayed by acids or bases.

15. I'm neither here nor there, so you see?

Neither an acid nor a base can I be.

Hint: I don't react with water to form acid or base.

16. I reduce what dissolves, that's my role?

By adding more ions, I take control.

When two share the same in a single pot,

Hint: I'm common but cause less of me to dissolve.

17. I split myself into two tiny parts?

H+ and OH–, with balanced hearts.

Even pure, I make ions without end,

Hint: Water can ionize itself, even when alone.

18. I'm the product of ions, so small yet clear?

A constant that's true, always near.

In water, I show what's hidden inside,

Hint: I measure the ion strength in pure H2O.

19. I carry the proton, I'm key to the pH?

In acidic solutions, I'm on a fast dash.

Three waters hold me, tight and strong,

Hint: I'm an H+ surrounded by water.

20. I carry a negative charge, can you see?

In bases, I'm as common as can be.

I bond with metals or float on my own,

Hint: I'm the reason basic solutions feel slippery.

21. I measure how much an acid will break,

In water, dissociation's at stake.

A value small or large defines my might,

Hint: I tell how much an acid likes to give up H+.

22. For bases, I show how much they'll split,

When in water, they don't always commit.

A number reveals their dissociation fate,

Hint: Bases dissociate, and I measure how well.

23. When salt meets water, it reacts just right,

To make it acidic or basic, in this fight.

I change the pH, just with a mix,

Hint: Salts alter water's pH when dissolved.

24. I've got more than one H+ to give,

I donate in steps, so I can live.

From sulfuric to phosphoric, I'm quite the case,

Hint: I release more than one proton, step by step.

25. With just one H+ I'm ready to share,

I lose it quickly, without much care.

I'm simple and quick, no need to debate,

Hint: Only one proton to give away.

26. Two protons I give, but not all at once,

I take my time, I'm no dunce.

First one goes, then the next,

Hint: I release two H+, one after the other.

27. I measure the strength of acids and bases,

Look to me when adjusting pH spaces.

Small values mean I'm strong, large means I'm weak,

Hint: Think logarithmic, I'm a sign of strength.

28. I change my color when the time is right,

In acids or bases, I show the fight.

From pink to orange, I help you see,

Hint: I signal when a reaction reaches the right point.

29. When acid and base meet face to face,

They cancel each other, in the neutral space.

Water and salt are the only remains,

Hint: The result is neither acidic nor basic.

30. At this stage, the titration is done,

The balance between acid and base has won.

No excess remains, both are neutralized,

Hint: It's where equal amounts of acid and base meet.

31. I come close to the point of equivalence,

A color shift tells me, with confidence.

Though I'm not exact, I still show the way,

Hint: The color change tells you I'm near the goal.

32. I keep the balance, like nature's glue,

In blood and cells, I hold pH true.

Without me, chaos would reign,

Hint: I stabilize pH in the body.

33. I fall from the sky with a sour touch,

Pollution causes me, it's too much.

I harm forests and lakes where I land,

Hint: I come from polluted air and harm nature.

34. I contain oxygen in my molecular frame,

Sulfur, phosphorus, and nitrogen, some of my fame.

I'm not binary, there's more to my form,

Hint: I contain oxygen along with H and another element.

35. I'm made of just two, no more in my name,

A hydrogen and a nonmetal, that's my game.

Simple and small, but I pack a punch,

Hint: I only have two elements in my formula.

36. When my bond is weak, I easily part,

The more polar, the easier to start.

I'm strong when my bond breaks right away,

Hint: Bond strength and polarity decide how strong I am.

37. I carry two charges, positive and negative too,

Ampholytes like me can do what few do.

In acids or bases, I can survive,

Hint: I carry both charges in the same molecule.

38. At this spot, I am neutral and calm,

No charge at all, I'm in perfect balm.

For proteins and molecules, this point is key,

Hint: I'm the pH where the molecule has no net charge.

39. When two come together, sharing a pair,

They form a bond and make a new lair.

Electron donors and takers, they find their match,

Hint: A product of two sharing electrons.

40. I speed things up, but I don't disappear,

Acids and bases, I hold near.

In reactions, I make things flow,

Hint: I make reactions faster without changing myself.

41. I'm stronger than sulfuric, a real force to fear,

Fluorine helps make me, powerful and clear.

I'm off the charts, with my proton might,

Hint: Think beyond normal acids; I'm much stronger.

42. I'm a metal that plays both roles,

In acidic or basic, I can control.

My ions shift between acid and base,

Hint: Some metals can switch sides.

43. I can donate or accept, it's true,

In acids or bases, I'll work with you.

I'm flexible in the proton game,

Hint: I can either give or take H+.

44. When I dissolve, my pH is low,

Though I come from a neutral show.

I make things sour, that's my fame,

Hint: I'm formed from a strong acid and a weak base.

45. I come from reactions, but don't stay neutral,

I tilt toward base, in ways quite crucial.

Hint: I come from a strong base and a weak acid.

46. I combine with water to make things sour,

Acidic in nature, I have the power.

Which hydride am I, with this trait,

Hint: I turn into an acid when mixed with water.

47. In water, I form a basic brew,

OH– appears when I'm with you.

I'm from a metal, strong and sure,

Hint: I form a base when mixed with water.

48. When acids and bases meet and greet,

I'm the energy they release as heat.

Exothermic in nature, I make things warm,

Hint: I'm the warmth from acid-base reactions.

49. Water surrounds, molecules dance,

Hydration's the process, given a chance.

Who am I, helping dissolve with grace,

Hint: I involve water interacting with ions.

50. Electricity flows when I'm around,

Ions move in a charged playground.

In acids and bases, I shine bright,

Hint: I measure how well solutions let charges move.

Chapter 6
Redox Reactions

1. I lose electrons and rise in state,

My process is vital to redox fate.

My charge grows higher, I cannot deny,

Hint: Loss makes me stronger, not weaker.

2. I gain electrons, it's quite the boon,

In the redox dance, I change my tune.

My charge drops lower, but don't feel bad,

Hint: I take what others give away.

3. I help others lose, while I gain a friend,

Electrons come to me, in the end.

I push oxidation, though I'm not so mean,

Hint: I cause others to lose, but I gain in return.

4. I give away electrons with ease,

Helping others to decrease.

I push reduction, while I'm left dry,

Hint: I help others gain, while I lose my share.

5. Electrons flow from one to another,

Oxidation and reduction, a twin brother.

A balance is struck, in this exchange,

Hint: It's a process of give and take with electrons.

6. A value I hold, to track the charge,

In redox reactions, I'm quite large.

I show how atoms shift and sway,

Hint: I represent the charge an atom would have if electrons were

completely transferred.

7. One species does both, oxidize and reduce,

In one reaction, I'm let loose.

Splitting into different states,

Hint: I undergo both oxidation and reduction simultaneously.

8. Two species combine, in opposite fates,

To form one in between, what awaits?

I'm the reverse of disproportionation's way,

Hint: It's like opposite parts finding middle ground.

9. I split the task into halves, you see,

One for oxidation, one for me.

We balance electrons to make things right,

Hint: I divide the reaction to balance each side.

10. I convert chemical energy to flow,

Electricity through wires will go.

Anode and cathode make up my frame,

Hint: I'm a device where redox reactions power electricity.

11. I work without power, spontaneous and free,

Converting chemicals to electricity.

Two half-cells make me complete,

Hint: I generate electricity on my own.

12. With external power, I make reactions go,

Driving change where they're too slow.

Splitting compounds with electric might,

Hint: I need a push to make reactions happen.

13. I'm the place where oxidation occurs,

Electrons leave, without much slur.

I'm positive in electrolytic play,

Hint: I'm where electrons are lost.

14. Reduction takes place, I gain electrons here,

In any redox cell, it's crystal clear.

I'm negative in galvanic shine,

Hint: I'm where electrons come and settle.

15. I balance the charge, between two halves,

Letting ions flow as the reaction laughs.

I connect the cells, but don't carry charge,

Hint: I allow ions to move while keeping electrons separate.

16. I move in reactions, from here to there,

Carrying charge, making the air.

I'm key to redox, it's clear to see,

Hint: I jump from one atom to another.

17. I'm the number assigned to an atom's fate,

Showing gain or loss in the redox gate.

A guide to charge in any reaction,

Hint: I help track how oxidized or reduced an atom is.

18. At standard conditions, I show my might,

The power of reduction, is in my sight.

Measured against hydrogen, I stand tall,

Hint: I compare everything to hydrogen's zero.

19. I adjust the voltage, based on ion flow,

Changing conditions, my numbers grow.

Predicting potential with concentrations in mind,

Hint: I calculate potentials based on the number of ions.

20. I rank the players, who's strong and who's not,

In terms of reduction, I show the plot.

From lithium to gold, I set the tone,

Hint: I rank elements by how easily they reduce.

21. I'm another name for galvanic might,

Where chemical reactions make light.

Spontaneous energy, I don't require a shove,

Hint: I'm the cell that makes electricity without external power.

22. I split compounds with electric shock,

Turning liquid into a reactive block.

Water or salts, I break with care,

Hint: I break down compounds using electricity.

23. I tell you how much substance will change,

With charge in reactions, it's no longer strange.

Proportional to current, time, and mass,

Hint: I tell you how much mass reacts based on the charge.

24. I measure the urge of a half-cell's drive,

To gain or lose electrons, and thrive.

Compared to hydrogen, I take my stand,

Hint: I predict how much a half-cell wants to gain or lose electrons.

25. I'm the voltage between two reacting sites,

The difference in drive creates electric lights.

Add me up for the total score,

Hint: I'm the total voltage of an electrochemical cell.

26. Extra energy is what I need,

To drive the reaction, beyond my speed.

I'm the excess, the cost to push,

Hint: I'm the additional voltage required for a reaction to proceed faster.

27. I eat away metals, slowly but sure,

Water and oxygen help me endure.

Rust is my fame, but I do much more,

Hint: I'm the unwanted reaction that destroys metals over time.

28. I form red flakes, when water's near,

Oxygen helps me, that much is clear.

I weaken structures, make them fall,

Hint: I make iron corrode in damp conditions.

29. I coat surfaces with metal shine,

A thin, protective layer, so fine.

Using electricity, I stick on tight,

Hint: I deposit metal layers using electrolysis.

30. I use chemicals to produce energy clean,

Hydrogen and oxygen are part of my scene.

No combustion here, just a steady flow,

Hint: I generate electricity from fuel without burning it.

31. A future of fuel, light as a breeze,

I promise clean energy, with no CO2 keys.

Hydrogen's my star, the cleanest by far,

Hint: I focus on hydrogen as a green fuel for the future.

32. I find the unknown with color's touch,

Redox reactions reveal so much.

Measuring the change in oxidation states,

Hint: I use redox to find how much of a substance is present.

33. Instead of colors, I use voltage's sway,

To find the point where redox reactions play.

No indicator needed, just a change in the read,

Hint: I track reactions by the potential change, not the color.

34. I measure oxygen in water so deep,

Using redox reactions that I keep.

To determine if life can survive or not,

Hint: I check how much oxygen is dissolved in water, vital for life.

35. I use iodine as my colorful friend,

Tracking how much reacts in the end.

Starch shows when I've had enough,

Hint: Iodine and starch are my signature pair.

36. With purple permanganate, I shine so bright,

Oxidizing substances, in my purple light.

I turn from color to color in a blink,

Hint: A bright purple ion changes as it reacts.

37. Orange to green is my color change,

In oxidizing substances, I rearrange.

I help in finding concentrations clear,

Hint: I use an orange oxidizer that turns green as it reacts.

38. I measure the oxygen needed to decay,

Organic matter in water, night or day.

I show how pollution affects life's beat,

Hint: I show how much oxygen organisms need to break down waste.

39. I measure how much oxygen is used,

To oxidize chemicals, never confused.

Not just for life, but for chemicals too,

Hint: I track oxygen used to break down chemicals, not just organic matter.

40. When reactions balance, and flows cease,

My state is reached, there's peace.

No more electrons jump from side to side,

Hint: I'm the point where electron flow stops because everything is balanced.

41. Use me once, then throw me away,

I store energy, but only for a day.

I'm not rechargeable, but I work fast,

Hint: I can't be recharged after use.

42. Use me once, then charge me again,

I'm a reusable battery friend.

I store energy and come back for more,

Hint: I can be recharged and used multiple times.

43. I power cars with a chemical spark,

With lead and acid in the dark.

I'm heavy and strong, and charge with care,

Hint: I'm a common car battery with lead in my name.

44. I store charge with a metal mix,

Nickel and cadmium in my battery fix.

I can be recharged but watch for decay,

Hint: I'm a rechargeable battery with nickel and cadmium at play.

45. Light and fast, I power your phone,

I'm the battery that's widely known.

I recharge quickly, with ions that flow,

Hint: I'm in your smartphone, and I recharge easily.

46. I'm cheap and common, but don't last long,

With zinc and carbon keeping me strong.

Disposable and light, I power small things,

Hint: I'm a common disposable battery, often found in small devices.

47. I last longer, with alkaline might,

In small devices, I'm quite the sight.

Disposable but strong, I power with grace,

Hint: I last longer than zinc-carbon, and you use me in remotes and toys.

48. From anode to cathode, I move with ease,

Carrying charge through wires like a breeze.

I'm the current that powers your day,

Hint: I'm the movement of negative charge through a circuit.

49. I'm the shorthand for reactions in a cell,

With vertical lines dividing quite well.

I show what's oxidized and what's not,

Hint: I'm a written representation of a galvanic cell.

50. I'm one side of the redox tale,

Oxidation or reduction without fail.

Paired with another, I make the whole,

Hint: I'm half of the electrochemical story.

51. I'm the reference for all electrode might,

Set at zero potential, pure and right.

Hydrogen bubbles at my side,

Hint: I'm the universal reference for measuring potentials.

52. I stay stable, so reactions can play,

Measuring potentials, I lead the way.

I don't change, no matter the time,

Hint: I'm an electrode that keeps constant potential.

53. I measure the strength of a redox urge,

To gain or lose electrons, I emerge.

Positive or negative, I tell the tale,

Hint: I predict how easily something will undergo a redox change.

54. I show the steps, oxidation through,

With potentials laid out in a view.

From one state to another, I help you see,

Hint: I map out reduction potentials for each oxidation state of an element.

55. I plot free energy against oxidation state,

Showing what's stable, what's up for debate.

My curve helps predict what's best,

Hint: I compare oxidation states to see which one's most stable.

56. I show the balance between redox and pH,

Telling you what will corrode or break.

Water and metals, I predict their fate,

Hint: I'm a plot of pH vs. potential showing stability regions for

species in water.

57. Electrons jump between partners near,

Creating bonds that they hold dear.

I form a complex, with colors so bright,

Hint: I involve a transfer of charge between molecules, often

forming colored compounds.

58. I'm a pair of states, oxidized and reduced,

Always together, never confused.

One loses, the other gains,

Hint: I'm the two forms of the same species in a redox reaction.

59. I make metals lose their shiny coat,

Turning them to oxide, no need for a vote.

I'm rust's cousin, and tarnish too,

Hint: I'm the process that makes metals corrode or rust.

60. I turn ores to metal, shiny and bright,

With electrons gained in this redox fight.

From oxide or ion to pure metal form,

Hint: I'm the opposite of oxidation, bringing metal to life.

61. I measure the push of electrons in flow,

Determining if reactions will go.

I'm the voltage at each electrode part,

Hint: I show how likely an electrode is to gain or lose electrons.

62. In cars, I clean the exhaust so well,

Using redox reactions, can you tell?

I turn harmful gases into safer air,

Hint: I reduce pollution from car exhaust.

63. I'm a disinfectant and a bleaching friend,

But in redox reactions, I don't pretend.

I can oxidize or reduce, it's true,

Hint: I can act as both an oxidizing and reducing agent.

64. I disinfect water and bleach clothes white,

But in redox reactions, I'm a powerful sight.

I oxidize with ease, leaving nothing behind,

Hint: I'm used in swimming pools and can kill bacteria.

65. I'm the most electronegative of all,

In redox reactions, I stand tall.

I pull electrons with all my might,

Hint: I'm the most reactive halogen.

66. In one reaction, I change both ways,

Oxidizing and reducing without delay.

Same element, two fates it meets,

Hint: I involve the same substance being both oxidized and reduced.

67. I'm the reference value in volts so neat,

Helping predict if reactions meet.

At standard conditions, I'm the key,

Hint: I'm the standard voltage for comparing redox reactions.

68. I go round and round, no end in sight,

Oxidation, reduction in cyclic might.

I'm used in processes that never stop,

Hint: I'm a continuous loop of redox reactions.

69. Two different metals in contact so close,

One will corrode more than most.

Through a redox process, one decays,

Hint: I involve the degradation of metals when two different ones touch in an electrolyte.

70. In plants, I split water to make food,

A redox reaction that sets the mood.

Oxygen's released, and sugars made,

Hint: I happen in plants and provide them with energy.

71. I break down glucose, releasing power,

A redox process every hour.

Electrons transfer, energy flows,

Hint: I release energy by breaking down food in cells.

72. I turn sugars red or blue,

Depending on what they do.

I'm a redox test for aldehydes,

Hint: I'm a test that shows if a sugar can be oxidized.

73. I'm a redox reaction, that's for sure,

Testing sugars in a mixture pure.

I go from blue to red when sugar's near,

Hint: I'm a test that turns red in the presence of reducing sugars.

74. I eat away at iron with air,

Water speeds me without care.

I'm redox in action, slow but sure,

Hint: I'm what happens to iron when exposed to oxygen and moisture.

75. I protect metals with a layer so thin,

Stopping redox reactions from within.

I keep corrosion at bay with ease,

Hint: I'm a protective oxide layer that forms on metals.

76. I use energy from the sun to play,

Hint: I use sunlight to catalyze redox reactions.

77. With light, I speed redox reactions bright,

Helping chemicals change in the light.

I use energy from the sun to play.

Hint: I use sunlight to catalyze redox reactions.

78. I'm the first battery, made of metal and brine,

Creating voltage with a design so fine.

I stack plates in a careful pile.

Hint: I'm the earliest form of a battery.

79. I generate power with hydrogen pure,

A redox reaction clean and sure.

I produce water as I work all day.

Hint: I use hydrogen and oxygen to create electricity with water as the by-product.

80. I take electrons in redox play,

Helping reactions go their way.

I'm reduced as I gain more.

Hint: I'm the species that gets reduced by gaining electrons.

81. I give electrons to my friend,

Helping redox reactions to the end.

I'm oxidized as I let them go.

Hint: I lose electrons in redox reactions.

82. I'm the voltage across a cell,

Measuring how reactions swell.

At standard conditions, I tell you the score.

Hint: I measure the potential difference under standard conditions.

83. In purple hues, I oxidize strong,

Changing colors as I go along.

I'm powerful, turning things to ash.

Hint: I'm a purple ion often used in redox titrations.

84. I slow down rust, protecting the day,

In redox reactions, I keep decay at bay.

I'm added to metals to keep them bright.

Hint: I'm a chemical that prevents rusting of metals.

85. I calculate potentials at any rate,

Changing with concentration's fate.

I modify voltages for real-life use.

Hint: I adjust cell potential based on ion concentration.

86. I split water into two gases bright,

Oxygen and hydrogen, a powerful sight.

Using redox and electric power.

Hint: I separate water into its two components using electricity.

87. I protect metals from corroding away,

By sacrificing myself day by day.

I oxidize first, saving my friend.

Hint: I protect metals like iron by corroding instead.

88. I convert fuel to energy with care,

Using redox reactions everywhere.

With hydrogen, I power up right.

Hint: I produce electricity from fuels like hydrogen.

89. I coat metals with a shiny layer,

Using redox to make things fairer.

I make objects gleam and glow.

Hint: I use electricity to deposit metal ions on a surface.

90. I turn silver black with time,

Through redox reactions in a grime.

It's sulfur's touch that makes me show.

Hint: I result from a reaction between silver and sulfur compounds in the air.

91. In cells, I break down sugars sweet,

Releasing energy with every beat.

Oxygen helps me in my task.

Hint: I'm how your body gets energy from sugar.

92. I reduce metals, making them pure,

With hydrogen gas, my process is sure.

I leave no trace, no oxide to see.

Hint: I'm a gas that helps remove oxygen from metal ores.

93. I light the sky with colors bright,

Using redox reactions in flight.

With metals and salts, I make a spark.

Hint: I cause vibrant colors through oxidation and reduction.

94. In orange hues, I take electrons with pride,

Helping organic compounds oxidize wide.

I change color as I oxidize.

Hint: I'm often used in titrations and reactions where alcohols become aldehydes.

95. I store energy in a chemical way,
With zinc and copper in an electrolyte play.

Electrons flow, and power is born.

Hint: I'm a simple battery using two different metals in an acid solution.

96. I protect aluminum with color and might,

Oxidizing its surface in an electric light.

The layer I form is thick and strong.

Hint: I create a corrosion-resistant layer on aluminum.

97. I change color to let you know,
When the redox reaction is about to show.

From one hue to another, I do my part.

Hint: I signal the end of a redox titration by changing color.

98. I'm a silver mirror, shiny and clean,
Testing aldehydes with a brilliant sheen.

I oxidize sugars, leaving silver behind.

Hint: I'm a test for aldehydes, and I leave a metallic mirror on glass.

Chapter 7

Coordination Compounds

1. A bond so strong, yet often light,

I hold the complex, day and night.

What's this connection that I bring,

In the dance of chemistry, I sing?

Hint: This bond is formed between a metal and a ligand.

2. I measure how strong a bond can be,

With constants that tell you, come and see.

What am I, in this stability game,

Determining which complex will gain fame?

Hint: My value indicates the stability of a coordination complex

in solution.

3. One step at a time, I measure the flow,

Of how ligands attach, as you surely know.

What's my name in this gradual climb,

Quantifying binding in a rhythmic rhyme?

Hint: I track stability as each ligand binds one by one.

4. In a grand equation, I show the whole,

How strong is the bond? That's my goal.

What's my title in this constant affair,

Summing up stability with utmost care?

Hint: I provide a single constant for the entire complex formation.

5. With a ring so large, I embrace my kin,

Holding metals tight, I draw them in.

What's my type, in this cyclic embrace,

Where stability finds its rightful place?

Hint: I form large ring structures that strongly bind metal ions.

6. In my circular arms, stability thrives,

Metal ions fit, and bonding arrives.

What's this phenomenon that makes me strong,

In coordination chemistry, where I belong?

Hint: This effect describes the enhanced stability of macrocyclic ligands compared to acyclic ones.

7. In the center of action, I proudly reside,

With colorful properties, I cannot hide.

What am I, in this dynamic realm,

Where transition metals take the helm?

Hint: These complexes involve d-block elements and often display vibrant colors.

8. In living systems, I play my role,

With metals and ligands, I make them whole.

What's this field where I contribute,

To life's processes, I take root?

Hint: I study how metal complexes interact in biological systems.

9. In a world of acids and bases, I reside,

With hardness and softness, I take my stride.

What's my theory in this chemical dance,

Guiding reactions with a glance?

Hint: I categorize acids and bases based on their reactivity and properties.

10. With carbon and metal, I share a bond,

In my structure, the two respond.

What's my type in this coordination spree,

Where carbonyls bond so elegantly?

Hint: I involve carbon monoxide as a ligand to a metal center.

11. I share my strength in a two-way street,

With π-back donation, I'm hard to beat.

What's this concept in bonding's embrace,

Where ligands help in a molecular race?

Hint: This interaction enhances the stability of metal-ligand bonds through π interactions.

12. With my empty orbitals, I reach for the metal,

Accepting electrons, I'm never a settle.

What's my role in this bonding tale,

Where my strength helps complexes prevail?

Hint: I accept electron density back from the metal through π-back bonding.

13. In my world, I give and share,

Donating electrons with utmost care.

What's my type in this bonding spree,

Where I form bonds quite readily?

Hint: I donate electron pairs to the metal center.

14. In my realm, electrons jump with glee,

From ligands to metal, can you see?

What's this process that gives me flair,

With energy transitions filling the air?

Hint: This involves an electron moving between the metal and ligand, affecting color.

15. In every bond, I take a stance,

My charge determines how I dance.

What's my role in this redox play,

Where oxidation numbers hold sway?

Hint: I indicate the charge of the metal in a coordination complex.

16. With names so tricky, I follow the rules,

Of ligands and metals, I'm one of the tools.

What's my task in this chemical game,

To describe compounds and make them famous?

Hint: I provide systematic names for coordination complexes based on IUPAC conventions.

17. With prefixes and suffixes, I help define,

How to name ligands, in a structured line.

What's my role in this naming spree,

Creating clarity for all to see?

Hint: I guide you on how to name simple and complex ligands according to IUPAC rules.

18. With different ways to hold on tight,

I bind in manners that feel just right.

What's my name in this binding lore,

With modes that offer so much more?

Hint: I describe how ligands can attach to the central metal, such as bidentate or tridentate.

19. With three points of contact, I hold on strong,

In the world of coordination, I belong.

What am I, in this bonding affair,

Securing my metal with utmost care?

Hint: I'm a ligand that binds through three donor atoms.

20. In a field of ions, I make my case,

Explaining colors and shapes in space.

What's my theory in this electronic sphere,

Where splitting and energies appear?

Hint: I describe how ligands influence the energy levels of d-orbitals in metal complexes.

21. With my advanced view, I take a leap,

Understanding bonding in a depth that's deep.

What's my theory in this chemical scene,

Where ligands and metals work as a team?

Hint: I build upon crystal field theory, incorporating covalent character.

22. In the realm of metals, I change my form,

With ligands nearby, I start to transform.

What's my process in this energetic game,

Where orbital energies never stay the same?

Hint: I describe how d-orbitals split in the presence of ligands.

Chapter 8
Nuclear Chemistry

1. In atoms I dwell, unstable and wild,

I release energy, like a curious child.

What am I, with a glow that you see?

A process that changes, come learn it with me!

Hint: I'm the reason why some elements are bold, giving off rays, both bright and cold.

2. In a world of elements, I have a twin,

Same protons, different neutrons, let the fun begin!

What am I, with variations so neat,

Changing my weight, while keeping my beat?

Hint: Think of carbon; one has a weight, others differ slightly, but they still relate.

3. Time ticks away, I measure decay,

In a nuclear dance, I show you the way.

What am I, where quantity falls,

To half of what once was, through nature's calls?

Hint: If you start with 100, in time I'll show, how much is left when the minutes flow.

4. I split with a force, breaking apart,

Releasing great energy, a scientific art.

What am I, with nuclei that divide,

Creating more power, in a brilliant stride?

Hint: Think of splitting a big cake, it gives you a slice; more pieces for all, isn't that nice?

5. Two into one, a powerful blend,

The sun's fiery heart, on me you depend.

What am I, where light elements merge,

Creating energy, in a cosmic surge?

Hint: It's the process that fuels stars up high; when atoms collide, watch the sparks fly.

6. I'm a transformation, a change so profound,

An element's journey, where new forms are found.

What am I, with particles that flee,

In a dance of decay, come discover me?

Hint: Like a seed that grows into a tree, I change my form, just wait and see.

7. I'm the highest energy, invisible to sight,

Traveling far, with powerful might.

What am I, with waves so rare,

Penetrating deeply, I'm everywhere?

Hint: Not a ray of sunshine, but I can still glow; I'm a radiation type that can put on a show.

8. Small and mighty, I carry a charge,

Two protons, two neutrons, at large.

What am I, with a name that's bold,

Ejecting from nuclei, a sight to behold?

Hint: Think of helium's core, as I take my leave; a heavy little

particle, you won't believe.

9. I travel fast, a speedy little friend,

Electrons or positrons, on me you can depend.

What am I, with a charge that's light,

Flashing through matter, in day or night?

Hint: A tiny burst of energy, I dart and dash; when you see me

coming, you'll hear a splash.

10. I'm the silent partner, in nuclear play,

When I collide, elements sway.

What am I, changing your view,

Creating isotopes, both old and new?

Hint: I'm not charged like my kin, but I play a key role; in

reactors and labs, I help make you whole.

11. With a tick of the clock, I measure the past,

Organic remains, my knowledge is vast.

What am I, in a scientific race,

Determining ages with a touch of grace?

Hint: From trees to ancient bones, I'll tell you their time; just look for the carbon, in layers of grime.

12. I gather my atoms, a powerful crowd,
At a certain point, I'm strong and loud.
What am I, when fission can start,
Creating a chain that can tear things apart?
Hint: I'm the threshold you reach to ignite a grand blast; when I'm met, energy is unleashed fast!

13. A spark turns to fire, as I multiply,
One atom's split leads to more—oh my!
What am I, a series of thrills,
An endless cycle that powers our drills?
Hint: Like dominoes falling in a perfect line, watch me unfold as I cross the design.

14. In my heart, the atoms collide,
Generating power, with nowhere to hide.
What am I, where fission takes place,
Providing energy for the human race?
Hint: Think of a furnace, but one that's controlled; I harness the heat that the atoms unfold.

15. A heavy metal, I'm prized and rare,
Fuel for reactors, with strength to spare.

What am I, with isotopes in my core,

A player in energy, forever in store?

Hint: I'm found in the earth, in rocks, and in sand; my power fuels nations across every land.

16. I'm born from uranium, a crafty sort,

In nuclear circles, I hold a strong court.

What am I, with a number that's bold,

A fuel for reactors and weapons of old?

Hint: I'm named after a planet, so far from your sight; my potential is vast, whether day or night.

17. Invisible gas, I linger with ease,

From rocks in the ground, I drift on the breeze.

What am I, with dangers in tow,

A noble gas, but with a harmful glow?

Hint: I can sneak in your home, through cracks and the floor; be cautious with me, or I'll knock at your door.

18. I mimic the sun, where energy's born,

Fusing together, my power is sworn.

What am I, in the quest for clean light,

A reactor of hope, burning ever so bright?

Hint: I bring the stars closer, in a quest to find peace; combining small atoms, my energy won't cease.

19. Slow and steady, I drift through the mass,

With a gentle touch, I help reactions amass.

What am I, with a warmth that's low,

In nuclear fission, I help things to grow?

Hint: Think of a soft breeze that nudges the leaves; I'm the

friendly neutron, in which energy weaves.

20. In the heart of the atom, I play a sly role,

The missing mass gives the energy a goal.

What am I, where numbers don't meet,

Binding particles close, as they dance to the beat?

Hint: Like a puzzle that's missing a piece or two, I explain why

energy's greater than what you knew.

21. I hold them together, in a nuclear embrace,

Creating stability in the atomic space.

What am I, with a force so strong,

In the nucleus' song, where I truly belong?

Hint: Without me, the atoms would fall apart; I'm the glue of the

nucleus, holding each heart.

22. I'm a particle of light, dancing through air,

No mass on my shoulders, yet I'm everywhere.

What am I, in waves or in beams,

Carrying energy, fulfilling dreams?

Hint: Think of the sun's rays that warm your skin; I'm the light you can't see, where journeys begin.

23. With precision and care, I help heal and mend,
In a world of diseases, I'm the patient's friend.
What am I, using isotopes with grace,
Diagnosing and treating in this vital space?
Hint: I bring together science and health with a spark; watch as I light up the dark!

24. I follow the path, in science I shine,
Revealing the secrets, in tissues and time.
What am I, in experiments bold,
Helping researchers see what's untold?
Hint: I'm often radioactive, but don't be afraid; I'm the guide on the journey where knowledge is laid.

25. Measuring the dose, I keep things in check,
In the realm of radiation, I earn my respect.
What am I, with tools precise and true,
Ensuring safety, as I measure for you?
Hint: Like a scale for your health, I ensure that you're fine; I help gauge the risks, and keep them in line.

26. Leftover materials, no longer in use,
Contain my secrets, a dangerous muse.

What am I, in barrels and drums,

A challenge to manage, where danger comes?

Hint: Think of trash from a nuclear dance; handling me wisely is your best chance!

27. With a dose of precision, I heal and detect,

A fusion of science, I earn your respect.

What am I, combining health with a ray,

Guiding doctors in a powerful way?

Hint: I'm the magic potion that glows in the dark; in the world of medicine, I leave my mark.

28. A chain of transformations, one leads to the next,

In the world of isotopes, I'm quite complex.

What am I, as elements change and flow,

From one to another, as time moves slow?

Hint: Like a family tree with branches that spread; I show how one atom can change into lead.

29. I'm a threat in the shadows, unseen and sly,

With my silent presence, I can make you cry.

What am I, with risks that you must face,

A danger in handling, in this scientific space?

Hint: I lurk in the corners where radiation flows; to keep safe from me, knowledge is how it grows.

30. Click, click, I measure, alerting the keen,

Radiation levels, I help you glean.

What am I, with a sound so clear,

Detecting the dangers that might come near?

Hint: Think of a detective, with a sensitive ear; I reveal what's

hidden, so have no fear!

31. In the world of decay, I'm a unit of measure,

Counting disintegrations, I'm a scientific treasure.

What am I, quantifying the rate,

Of radioactive decay, with knowledge so great?

Hint: Named after a pioneer who shone in the light; I help

scientists see what's wrong and what's right.

32. With a name so bold, I measure the glow,

A legacy of science, in which I bestow.

What am I, named after two with great flair,

In the realm of radioactivity, I'm quite rare?

Hint: I'm a unit of activity, in the nuclear game; think of Marie's

brilliance, forever in fame.

33. In the world of dose, I'm your safety guide,

Measuring radiation that may come inside.

What am I, with a name so precise,

Quantifying the risks, I'm very nice!

Hint: Named after a physicist, wise and astute; I help keep you safe when dealing with loot.

34. I'm an isotope that's unstable and free,

Emitting radiation, just wait and see!

What am I, in a world so profound,

With properties that change when new forms abound?

Hint: I can be found in both nature and lab; some of my kinds can be quite the fab!

35. When I meet a nucleus, I take a new role,

Becoming part of it, achieving my goal.

What am I, with a quiet embrace,

Creating isotopes in the atomic space?

Hint: Like a gentle hug in the atomic fray; I help create new forms in my special way.

36. High temperatures rise, as nuclei fuse,

In the heart of the stars, I'm the one you choose.

What am I, where energy bursts forth,

Creating the heat that warms our Earth's girth?

Hint: Think of the sun, where my dance is grand; I'm the power of fusion, across every land.

37. I flow through the heart, keeping things chill,

Regulating temperature, with a scientific thrill.

What am I, vital for safety's embrace,

Ensuring the reactor stays in its place?

Hint: I'm the liquid or gas that keeps the heat low; in the nuclear world, I help things to flow.

38. After the split, I'm what's left behind,

A variety of elements, unique and unkind.

What am I, with traits that vary so much,

From isotopes to particles, I have quite the touch?

Hint: When atoms break apart, I take my stand; in the aftermath of fission, I'm what's planned.

39. I stand in your way, protecting your health,

Blocking the rays, guarding your wealth.

What am I, with layers that guard,

Against radiation that can hit hard?

Hint: Think of a wall, but one that's quite smart; I'm what keeps you safe, playing a vital part.

40. The study of forces that govern the core,

Exploring the mysteries, forever in store.

What am I, where particles collide,

Revealing the secrets of matter inside?

Hint: A branch of science that dives deep into fate; I unlock the universe's intricate gate.

41. In the realm of fission, I'm a crucial point,

Where reactions begin, to energy's joint.

What am I, when the balance is right,

Creating a chain in the nuclear night?

Hint: Think of a scale, where the weight is just so; too much or too little, and off we go!

42. I'm light and elusive, hardly seen at all,

Passing through matter, I slip and I crawl.

What am I, a ghostly particle's game,

With a role in the universe, never the same?

Hint: I dance through the cosmos without a care; I'm nearly massless, floating everywhere.

43. Without a nudge, I break all alone,

An atom's split that's all on its own.

What am I, with a sudden release,

Creating new elements, a natural piece?

Hint: Think of a pop, without warning or plan; I'm the fission that happens without any man.

44. In reactors I sit, holding energy tight,

With fission occurring, I'm crucial to might.

What am I, in a long slender line,

Providing the power, like a spine?

Hint: I'm the core of the reactor, filled with delight; without me, the fission would lose its light.

45. I manage the flow, keeping reactions in line,
Absorbing neutrons, making things fine.
What am I, with materials strong,
Balancing the power, where I belong?
Hint: Like a brake in a car, I help keep control; I'm essential for safety in the nuclear role.

46. With energy so high, I strip away charge,
Creating ions, in a process large.
What am I, transforming matter around,
As electrons are freed, and new states are found?
Hint: Think of a zap, that changes the game; I'm the process that electrifies, calling your name!

47. An electron's embrace, I pull it inside,
Changing my element, in a quantum tide.
What am I, where particles meet,
Creating a balance, so bittersweet?
Hint: Think of a dance where one partner is shy; I'm the capture that alters, making atoms comply.
48. I'm a transformation, with particles that race,
A neutron turns into something, finding its place.

What am I, as I change my form,

With an electron's release, a new norm?

Hint: Like a seed sprouting into a flower, I'm a change in the atom, revealing my power.

49. I'm the danger that lurks in the shadows of rays,

With harmful effects that can cause dismay.

What am I, with risks that you see,

A measure of danger in radiation's spree?

Hint: In the world of radiation, I carry the threat; understanding my levels is a wise bet.

50. A sequence of splits, where one sparks the next,

In the dance of decay, I'm complex.

What am I, in a nuclear embrace,

Creating vast energy, in this special space?

Hint: Like a row of dominoes falling in line, I generate power as I intertwine!

ORGANIC CHEMISTRY

Chapter 9
Concepts In Organic Chemistry

1. In bonds I play, holding charge tight,

The power to attract, I shine so bright.

From atoms to molecules, I define the pull,

In the realm of chemistry, I'm never dull.

Hint: I measure an atom's ability to attract electrons in a bond.

2. In my shared dance, I hold the weight,

One atom pulls harder, defining our fate.

With charge uneven, I create a scene,

In the world of chemistry, I'm in between.

Hint: It's a bond between atoms with different electronegativities, resulting in partial charges.

3. In symmetry we glide, evenly we share,

No charge distinction, we're a perfect pair.

In my calm embrace, no pull to be found,

In the world of molecules, I'm safe and sound.

Hint: It's a bond where electrons are shared equally between atoms.

4. In the dance of charge, I point the way,

Indicating polarity, come what may.

A vector I am, measuring the strife,

In the world of molecules, I show the life.

Hint: I quantify the polarity of a molecule based on charge separation.

5. In structures I play, not just one way,

Multiple forms, I lead the ballet.

Delocalizing electrons, I share the stage,

In the world of chemistry, I'm wise for my age.

Hint: I describe the phenomenon where molecules can be represented by two or more valid Lewis structures.

6. In my realm of electrons, I spread the cheer,

Sharing with neighbors, I hold them near.

Stability I bring to the aromatic ring,

In the world of chemistry, I'm a special thing.

Hint: I refer to the spread of electrons across adjacent atoms, increasing stability.

7. With a pull from afar, I influence fate,

Through bonds I extend, I resonate straight.

In the tug of electrons, I take my stance,

In the world of chemistry, I lead the dance.

Hint: I'm the electron-withdrawing or donating effect of substituents on a molecule.

8. In the overlap of bonds, I find my grace,

Electron donation gives stability a place.

Through sigma bonds, I spread my charm,

In the world of chemistry, I keep it warm.

Hint: I refer to the interaction between electrons in a sigma bond and an adjacent empty or partially filled p-orbital.

9. In a cyclic embrace, I bring delight,

With resonance and planarity, I shine bright.

A special stability, I hold so dear,

In the world of chemistry, I'm always near.

Hint: I describe the increased stability of aromatic compounds due to resonance.

10. In the confines of my circle, I feel the stress,

When angles are forced, I must confess.

With bonds that twist, I'm not quite at ease,

In the world of chemistry, I'm hard to please.

Hint: I refer to the increased energy and reactivity in cyclic compounds due to bond angle strain.

11. In the world of molecules, I'm subtle and shy,

A weak attraction, yet I can't deny.

From noble gases to long chains I span,

In the realm of bonding, I play my plan.

Hint: I include weak intermolecular forces that arise from temporary dipoles.

12. I'm a special attraction, strong yet fine,
Between hydrogen and something with a line.
With O, N, or F, I create a tie,
In the world of chemistry, I help compounds fly.
Hint: I'm a strong type of dipole-dipole interaction involving hydrogen atoms.

13. In a solvent's embrace, I help to dissolve,
With my interactions, problems I solve.
Surrounding ions or molecules with care,
In the world of solutions, I'm always there.
Hint: I refer to the process of solvent molecules surrounding solute particles.

14. In my intricate dance, I bond with pride,
With ligands and metals, I take my ride.
A complex I form, with bonds that align,
In the world of chemistry, our fates intertwine.
Hint: I involve the interaction of a central atom or ion with surrounding ligands.

15. With dots and lines, I sketch the scene,
Showing electrons and bonds in between.

A visual guide to molecular art,

In the world of chemistry, I play my part.

Hint: I'm a diagram that represents the valence electrons of atoms in a molecule.

16. In the realm of waves, I take my stand,

Explaining bonding through a quantum band.

With orbitals that mix, I show how they bond,

In the world of chemistry, my theories respond.

Hint: I describe the behavior of electrons in molecules through wave functions.

17. In bonds I reside, a classic view,

Explaining how atoms hold tightly, too.

With overlapping orbitals, I weave my tale,

In the world of chemistry, I will not fail.

Hint: I describe the formation of covalent bonds through orbital overlap.

18. In the blending of orbitals, I find my flair,

Creating new shapes with electrons to share.

From sp to sp^3, I define the game,

In the world of bonding, I earn my name.

Hint: I involve the mixing of atomic orbitals to form new hybrid orbitals.

19. In the first embrace, I take the lead,

A head-on connection, a strong bond indeed.

With overlapping orbitals, I hold it tight,

In the world of chemistry, I shine so bright.

Hint: I'm the type of bond formed by the direct overlap of orbitals.

20. I follow behind, a side-to-side sway,

In double and triple bonds, I make my play.

With orbitals aligned, I add to the show,

In the world of bonding, I help it flow.

Hint: I'm a bond formed by the lateral overlap of p orbitals.

21. In a ring of delight, I'm stable and neat,

With delocalized electrons, I can't be beat.

Following Hückel's rule, I take my stand,

In the world of chemistry, I'm perfectly planned.

Hint: I includes cyclic compounds with delocalized pi electrons that follow specific stability criteria.

22. In a chain of carbons, I add a twist,

With halogens attached, I can't be missed.

In reactions, I play, both diverse and wide,

In the world of chemistry, I take pride.

Hint: I'm an organic compound containing a carbon-halogen bond.

23. In my chain of carbons, I have a friend,
With sulfur attached, our bond won't bend.
With a dash of reactivity, I join the play,
In the world of chemistry, I'm here to stay.
Hint: I'm an organic compound containing a sulfur atom bonded to alkyl groups.

24. With a nitrogen base, I'm part of the crew,
In my chain of carbons, I add a hue.
In reactions and bonds, I'm often sought,
In the world of chemistry, I'm highly thought.
Hint: I'm an organic compound where a nitrogen atom is bonded to one or more alkyl groups.

25. With metals and carbons, I form my kin,
Creating reactions that spark from within.
In the realm of synthesis, I have my say,
In the world of chemistry, I pave the way.
Hint: I refer to compounds containing a metal atom bonded to a carbon atom of an organic group.

26. With strategy and thought, I lay the ground,
Charting the course for the products I've found.

In the lab, I guide each careful move,

In the world of chemistry, I help improve.

Hint: I'm the process of designing a synthetic pathway for creating compounds.

27. In the lab, I change from one to the next,

Transforming compounds with care and context.

A versatile player, I shift with grace,

In the world of chemistry, I find my place.

Hint: I involve the transformation of one functional group into another.

28. In the heart of reactions, I seek to explain,

The steps and pathways that lead to gain.

With logic and insight, I delve into schemes,

In the world of chemistry, I fulfill dreams.

Hint: I provide a logical framework for understanding how reactions proceed.

29. In the realm of science, I focus on rate,

With concentration and time, I calculate.

In the world of reactions, I measure the flow,

In the world of chemistry, I help you know.

Hint: I study the speed of reactions and the factors that influence it.

30. In the heart of equilibrium, I find my way,

Balancing reactions in a delicate play.

With shifts and changes, I keep it right,

In the world of chemistry, I shed my light.

Hint: I describe the state where the rate of forward reaction equals the rate of the reverse reaction.

Chapter 10

Hydrocarbons

1. In a world of single bonds, I stand tall,

Hydrogens and carbons make up my call.

I'm saturated, stable, and not too reactive,

A straight path I follow, simple and attractive.

Hint: Think of the simplest chains.

2. With one double bond, I twist and I turn,

Reacting with ease, it's my way to learn.

Add more atoms, watch me grow,

In this game of bonds, I steal the show.

Hint: I have a little twist in my structure.

3. Triple bonds hold me in a tight embrace,

A lively reaction, I pick up the pace.

I'm more reactive, with a sharper tone,

In the dance of bonds, I stand alone.

Hint: Think of a bond that triples in strength.

4. In a ring of carbon, my magic is found,

Resonance and stability, my beauty astound.

Delocalized electrons give me a name,

In fragrant compounds, I play the game.

Hint: I'm known for my sweet-smelling presence.

5. All bonds are single, my structure's complete,

With maximum hydrogens, I'm hard to beat.

No more room for the double or triple,

In this hydrocarbon world, I'm no little ripple.

Hint: Full of hydrogens, I am.

6. With double bonds, I crave a bit more,

Less hydrogen means I'm ready to explore.

Reactivity is my middle name,

In the game of bonds, I'll stake my claim.

Hint: I can bond in ways that others can't.

7. Same atoms arranged in a different way,

In the world of compounds, I'm here to play.

Though my makeup's the same, my face will change,

In the realm of chemistry, I'm quite strange.

Hint: Think of different shapes from the same parts.

8. In a puzzle of atoms, I find my delight,

Same formula, yet I hold a different sight.

With bonds re-arranged, my form I'll show,

Two structures from one, now you know.

Hint: I change my layout but not my content.

9. In the dance of molecules, I take my stance,

Cis and trans, it's all about the chance.

Spatial arrangements make me unique,

In the world of isomers, I'm what you seek.

10. On the same side, we love to align,

With a twist in our bond, our beauty will shine.

When the groups are close, our charm is revealed,

In the realm of isomers, my fate is sealed.

Hint: I prefer to stay together, close and near.

11. Across the way, I stand so proud,

With groups on opposite sides, I draw a crowd.

In my structure, symmetry plays its part,

In the world of isomers, I'm quite the art.

Hint: Think of separation in a bond.

12. A string of carbons, connected with might,

With hydrogens tagging along, oh what a sight!

In simplicity lies my strength and grace,

In the world of molecules, I find my place.

Hint: I'm a backbone in many compounds.

13. In a straight line, I stretch so far,

No bends or twists, just like a star.

With each carbon in a row, I proudly stand,

In the game of alkanes, I'm the simplest brand.

Hint: Think of a ruler—long and straight.

14. From a straight path, I take a turn,

With branches of carbon, it's my way to learn.

More complex than linear, I like to explore,

In the world of hydrocarbons, I'm never a bore.

Hint: I have offshoots from my main trunk.

15. In a circle, I twist, with bonds all around,

Saturated and stable, in cycles I'm found.

No ends to my chain, I loop and I spin,

In the dance of chemistry, I'm ready to win.

Hint: Think of a ring without a beginning or end.

16. Multiple rings are where I reside,

In the world of complex, I take great pride.

Stability through structure, my beauty displayed,

In the chemistry arena, I'm not easily swayed.

Hint: I'm made up of more than one ring.

17. Not in a ring, but chains I do boast,

With open chains of carbons, I'm the host.

Saturated or not, I cover the scene,

In the realm of hydrocarbons, I'm quite the machine.

Hint: I prefer my structure to be open and free.

18. In my cyclic realm, stability's key,

With delocalized electrons, I dance with glee.

My ring is special, with charm to impress,

In the world of chemistry, I'm truly blessed.

Hint: I have a special quality that brings stability.

19. A hexagon I am, with bonds that resonate,

In fragrant compounds, I hold a great fate.

Six carbons in harmony, with hydrogens to spare,

In the world of rings, I'm beyond compare.

Hint: I'm the simplest aromatic compound.

20. A methyl group here, with a benzene in tow,

In solvents and fragrances, I put on a show.

Not just a hydrocarbon, I have a flair,

In the world of chemistry, I'm always there.

Hint: I'm closely related to benzene, but with a twist.

21. With two methyl groups, I take my place,

In aromatic rings, I find my grace.

Three forms of me, each unique in design,

In solvents and scents, I truly shine.

Hint: I'm a close relative of benzene, but not alone.

22. In two rings fused, I dance with delight,

A solid at room temp, but sublime in flight.

In mothballs and fragrance, my presence is felt,

In the aromatic family, my charm is heartfelt.

Hint: Think of a compound known for keeping pests at bay.

23. In multiple rings, my structure is bold,

Complex and stable, with stories untold.

From smoke and soot, my origins trace,

In the world of chemistry, I find my place.

Hint: I'm composed of several interconnected aromatic rings.

24. A reaction with oxygen, I burst into flames,

Producing heat and light, playing my games.

From fuels to energy, I pave the way,

In the dance of molecules, I lead the sway.

Hint: I'm a common process for releasing energy.

25. When I lose some hydrogen, my fate is sealed,

With oxygen joining, new products revealed.

In this transformative dance, I'll change my face,

In reactions of energy, I find my place.

Hint: It's about losing hydrogen while gaining oxygen.

26. I break down long chains to make them small,

Creating alkenes from the big and tall.

In refining processes, I play my role,

Transforming hydrocarbons, that's my goal.

Hint: I help in turning heavy fractions into lighter ones.

27. From simple parts, I build something new,

Creating complex chains, it's what I do.

With reactions and steps, I'll craft my design,

In the lab of chemistry, my art will shine.

Hint: I'm about building compounds from simpler ones.

28. With a catalyst's help, I change my face,

Transforming hydrocarbons with elegant grace.

Boosting octane, I'm prized in the game,

In the world of fuels, I make my name.

Hint: I enhance fuels for better combustion performance.

29. A branch of carbon, I add some flair,
In the world of hydrocarbons, I'm everywhere.

From chains to rings, I change the tune,

In chemical names, I'm often a boon.

Hint: I'm a fragment that can modify structures.

30. In open chains, I wander and roam,
No rings to confine me, I find my home.

Saturated or not, my structure's defined,

In the realm of molecules, I'm one of a kind.

Hint: I prefer my structure without any cycles.

Chapter 11
Functional Groups

1. With an -OH group, I stand apart,

In drinks and solvents, I play a part.

From ethanol to spirits, I have my way,

In the realm of chemistry, I'm here to stay.

Hint: I'm known for my role in beverages.

2. Two groups I link with an oxygen in sight,

In the world of solvents, I'm quite light.

Not too reactive, but useful indeed,

In organic chemistry, I take the lead.

Hint: Think of me as a bridge between two groups.

3. At the end of my chain, a carbonyl's near,

With a hydrogen close, I draw you near.

In scents and flavors, my charm is found,

In the world of carbon, I'm renowned.

Hint: I often contribute to pleasant aromas.

4. In the middle of a chain, my carbonyl lies,

With two carbon neighbors, I claim my prize.

From sweet to savory, I flavor the dish,

In organic chemistry, I fulfill a wish.

Hint: I'm a carbonyl that loves company on both sides.

5. With -COOH, I'm sour, it's true,

In vinegar and citrus, I come through.

A blend of acid and carbon, I stand tall,

In the realm of compounds, I'm used in all.

Hint: I bring acidity to many natural substances.

6. With a fruity scent, I make you smile,

From alcohol and acid, I form in style.

In perfumes and flavors, I play my part,

In the world of esters, I'm a work of art.

Hint: I often contribute to the aroma of fruits.

7. With nitrogen bound, I'm often quite kind,

In proteins and bases, I'm easy to find.

From ammonia I came, with a unique flair,

In the realm of molecules, I'm everywhere.

Hint: I'm a nitrogen-containing compound that acts like a base.

8. A carbonyl with a nitrogen in place,

In proteins and peptides, I find my grace.

From acids I form, in bonds so tight,

In the world of chemistry, I'm a delight.

Hint: I'm related to both carboxylic acids and amines.

9. With a -SH group, I bring a strong smell,

In scents and chemicals, I do quite well.

Similar to alcohols, but with a twist,

In organic chemistry, I can't be missed.

Hint: I'm known for my characteristic odors.

10. With a carbon triple-bonded to nitrogen's might,

In plastics and fibers, I bring the light.

In the world of compounds, I stand quite alone,

In my group of friends, I'm well-known.

Hint: I'm often found in synthetic materials.

11. With a carbonyl and a chlorine in tow,

I'm reactive and useful, as you may know.

In making esters, I play my role,

In the realm of chemistry, I'm on a roll.

Hint: I'm often used in the synthesis of carboxylic acids.

12. Two acids combined, with water set free,

In a cyclic dance, that's how I'll be.

Reactive and versatile, I carry the load,

In the world of chemistry, I'm on the road.

Hint: I'm formed when two carboxylic acids lose water.

13. With a hydroxyl on my aromatic ring,

I'm known for my strength, and I often bring.

From antiseptics to plastics, I'm found all around,

In the world of compounds, I'm quite renowned.

Hint: I'm related to benzene but have a distinct functional group.

14. With a double bond between carbon and nitrogen,

In organic synthesis, I'm second to none.

From aldehydes I form, with a twist so fine,

In the realm of compounds, I gracefully shine.

Hint: I often appear in reactions involving aldehydes.

15. With a nitrogen in the scene, I take my place,

From ketones I form, with a different grace.

In reactions I thrive, as I change my face,

In the world of amines, I've found my space.

Hint: I'm related to imines but with an extra twist.

16. With a sulfur bond, I stand in the fray,

Two carbon chains linked in a special way.

In the world of compounds, I add a flair,

In fragrances and flavors, I'm often there.

Hint: Think of me as a cousin to ethers, but with sulfur.

17. Two sulfur atoms in a strong embrace,

In proteins and bonds, I find my place.

Stability and strength, that's what I bring,

In the world of chemistry, I'm an important thing.

Hint: I'm often involved in stabilizing protein structures.

18. With an O-O bond, I'm reactive and bold,

In bleaching and disinfecting, my stories are told.

Not just water, but with an extra flair,

In the chemistry game, I'm quite rare.

Hint: I'm known for my strong oxidizing properties.

19. In energy transfer, I play my part,

From ATP to DNA, I'm quite smart.

With oxygen and phosphorus, I form my chain,

In the world of biochemistry, I reign.

Hint: I'm vital for life and involved in energy storage.

20. With a -SO₃H group, I'm acidic and keen,

In detergents and dyes, I'm often seen.

My presence brings strength, in solutions I flow,

In the realm of acids, I put on a show.

Hint: I'm stronger than regular carboxylic acids.

21. Nitrogen with two oxygens in tow,

In explosives and dyes, I often show.

My strength and reactivity can't be denied,

In the world of chemistry, I take pride.

Hint: Think of a group known for its energetic compounds.

22. Two nitrogens linked by a double bond,

In vibrant dyes, my colors respond.

From yellow to red, I light up the scene,

In the chemistry world, I'm vibrant and keen.

Hint: I'm known for my brilliant colors in dyes.

23. With a C=O bond, I'm quite the sight,

In aldehydes and ketones, I bring the light.

In many reactions, I take the lead,

In the realm of compounds, I'm a vital seed.

Hint: I'm a functional group that appears in various compounds.

24. With -OH in my grasp, I stand so proud,

In alcohols and phenols, I draw a crowd.

Bringing properties that help me thrive,

In the world of molecules, I come alive.

Hint: I'm a common feature in many organic compounds.

25. An -O- link in a hydrocarbon chain,

In ethers I'm found, bringing joy to the gain.

With one side alkyl, I'm part of the show,

In the world of chemistry, I help compounds grow.

Hint: I'm often seen in ethers and related compounds.

26. With a double bond at the end of my chain,

In polymers and plastics, I stake my claim.

Flexible and strong, I'm quite the delight,

In the world of compounds, I'm a shining light.

Hint: I'm known for my role in creating synthetic materials.

27. With a three-carbon twist, I come into play,

In reactions and scents, I find my way.

From garlic to mustard, my presence is clear,

In the world of flavors, I bring good cheer.

Hint: I'm known for my connection to strong-smelling compounds.

28. In a ring with nitrogen, I take my stand,
A versatile compound, with uses so grand.

In synthesis, I play a vital role,

In the world of organic, I'm on a roll.

Hint: I'm used in reactions involving amines and acids.

29. With a halogen bonded, I'm ready to bond,
In the world of chemistry, I'm often fond.

From fluorine to iodine, I'm diverse and keen,

In organic synthesis, I can often be seen.

Hint: I'm a group that includes elements like chlorine and bromine.

30. In a cyclic dance, my bonds are unique,
With delocalized electrons, I'm strong, not weak.

In fragrances and flavors, I find my grace,

In the world of compounds, I hold my place.

Hint: I'm a structure commonly associated with benzene.

Chapter 12

Steriochemistry

1. In my structure, I hold a special place,

With four different groups, I set the pace.

Creating mirror images that can't be the same,

In the world of molecules, I play my game.

Hint: I'm a crucial point for chirality in compounds.

2. A twist in my structure makes me unique,

With left and right forms that can't truly meet.

In nature's designs, I often appear,

In the world of isomers, I bring cheer.

Hint: I'm all about those mirror images.

3. Two forms of me, mirror images in space,

Identical in properties, yet not a trace.

With different reactions, I take my stance,

In the world of stereochemistry, I lead the dance.

Hint: We're like left and right hands, but can't be superimposed.

4. Not quite mirror images, I take my form,

With different configurations, I'll weather the storm.

In the world of isomers, I hold my ground,

With varied properties, I can be found.

Hint: I differ in at least one, but not all, chiral centers.

5. A blend of enantiomers, equal in might,

In optical activity, I dim the light.

With no net rotation, I stand in between,

In the world of mixtures, I'm often seen.

Hint: I contain equal amounts of both chiral forms.

6. When I rotate light, I take the stage,

With chiral compounds, I'm all the rage.

Clockwise or counter, I show my flair,

In the realm of chemistry, I'm quite rare.

Hint: I'm a property of chiral substances affecting polarized light.

7. With a twist and a turn, I'm measured with care,

A ratio of angles that I gladly share.

In the world of optics, I help define,

In the study of chirality, I truly shine.

Hint: I help quantify the optical activity of compounds.

8. With the same formula, I differ in form,

In spatial arrangements, I weather the storm.

From enantiomers to diastereomers, I play,

In the world of isomers, I lead the way.

Hint: I share a molecular formula but differ in arrangement.

9. In a view from the front, I help you to see,

The rotation and bonds in a molecular spree.

Visualizing angles, I clarify the scene,

In the realm of structure, I'm a modeling queen.

Hint: I offer a way to look at bonds in 3D.

10. With lines and crosses, I lay out my plan,

Representing sugars, I'm part of the clan.

In vertical and horizontal, my groups will align,

In the study of stereochemistry, I'm quite fine.

Hint: I'm often used for carbohydrates and other compounds.

11. In a system so clear, I denote my place,

With priorities set, I define my face.

Clockwise or counter, I give you my name,

In stereochemistry, I play the game.

Hint: I'm a way to describe chiral centers distinctly.

12. In the world of alkenes, I find my rank,

With priority rules, I'll fill my tank.

For geometrical isomers, I help you decide,

In the realm of double bonds, I take pride.

Hint: I denote the arrangement around double bonds.

13. With a special bond, I hold my ground,

Creating stereoisomers, my traits abound.

In the world of chirality, I play my role,

In the dance of molecules, I'm on a stroll.

Hint: I'm synonymous with chiral centers in a compound.

14. With symmetry in form, I don't take a twist,

In mirror images, I can't be missed.

No chiral centers, I'm simple and plain,

In the world of isomers, I remain.

Hint: I look the same from all angles, no matter the view.

15. In my bonds, I twist and I turn,

With different shapes, there's so much to learn.

In the world of rotation, I play my part,

In molecular movement, I'm a work of art.

Hint: I'm about variations due to rotation around single bonds.

16. With planes of symmetry, I'm quite unique,

In the world of stereoisomers, I'm at my peak.

Despite having chiral centers, I'm not chiral at all,

In the realm of chemistry, I stand tall.

Hint: I have chiral centers but overall symmetry.

17. In separating enantiomers, I take my stance,

With methods and techniques, I enhance the chance.

Isolating the pairs, I help clarify,

In the world of chirality, I reach for the sky.

Hint: I'm the process of separating optical isomers.

18. In the realm of chirality, I find my home,

With mirror images, in pairs we roam.

Identical in formula, yet different in light,

In the dance of molecules, we take flight.

Hint: I'm synonymous with enantiomers.

19. In balance and order, I find my grace,

In shapes and forms, I hold my place.

In chiral compounds, I draw the line,

In the world of chemistry, I'm simply divine.

Hint: I'm often the reason why something is achiral.

20. In the art of creation, I lead the way,

Building chiral molecules, come what may.

With selectivity, I craft with delight,

In the world of chemistry, I shine bright.

Hint: I'm the method of creating one enantiomer over another.

21. With priorities set, I define my place,

In chiral centers, I bring my grace.

Clockwise or counter, I help to determine,

In the world of stereochemistry, I'm the firm one.

Hint: I'm used to assign R or S configurations.

22. In reactions I guide with an electronic touch,

The geometry and orbitals, I influence much.

In molecular interactions, I play a part,

In understanding structures, I'm quite smart.

Hint: I relate stereochemistry to electronic effects.

23. In my structure, symmetry seems to play,

But in reality, I'm led astray.

Not truly identical, yet I look the same,

In the realm of molecules, I make my claim.

Hint: I can appear symmetric without being truly so.

24. One substrate, one product, I make it clear,

In a specific way, I hold dear.

With stereochemistry guiding my fate,

In the dance of molecules, I navigate.

Hint: I yield products with a defined stereochemistry.

25. In reactions I favor, one path to embrace,

Creating one isomer, I set the pace.

With selectivity in bonds, I lead the way,

In the world of chemistry, I seize the day.

Hint: I prefer one stereoisomer over another.

26. In my bonds, I twist and I sway,

Exploring the shapes in a molecular ballet.

From staggered to eclipsed, I take my turn,

In the realm of structures, I help you learn.

Hint: I refer to the different spatial arrangements of a molecule.

27. In the world of cyclohexanes, I take my form,

With stability and comfort, I weather the storm.

No strain in my bonds, I sit up so tall,

In the realm of rings, I'm the best of them all.

Hint: I'm the most stable conformation of cyclohexane.

28. With my shape like a boat, I float in the fray

Less stable than a chair, I sway and I sway.

With torsional strain, I try to abide,

In the world of cyclohexanes, I take a ride.

Hint: I'm a less stable conformation of cyclohexane.

29. In cyclohexane, I make my choice,

With axial up and equatorial's voice.

In the dance of sterics, I find my place,

In the world of rings, I set the pace.

Hint: I refer to the two distinct orientations of substituents on a cyclohexane ring.

30. In a twist and a turn, I change my view,

From axial to equatorial, I make it true.

With bonds that rearrange in a graceful spin,

In the world of conformations, I always win.

Hint: I'm the process of shifting positions on a cyclohexane ring.

Chapter 13

Reaction Mechanisms

1. With electrons in hand, I'm ready to attack,

Seeking a positive charge, I'm on the right track.

In reactions I flourish, making bonds anew,

In the world of chemistry, I'm your clue.

Hint: I'm a species that donates electron pairs.

2. With a positive charge, I'm eager to gain,

Seeking electrons, it's part of my game.

In reactions I thrive, with nucleophiles in tow,

In the world of chemistry, I'm in the flow.

Hint: I'm a species that accepts electron pairs.

3. In the middle I stand, with energy high,

At the peak of the hill, I touch the sky.

In a fleeting moment, I'm here and then gone,

In the dance of reactions, I carry on.

Hint: I represent the highest energy point during a reaction.

4. In the reaction, I'm part of the scene,

Formed and then broken, I'm caught in between.

With bonds that shift, I play my part,

In the world of mechanisms, I'm a vital art.

Hint: I'm a transient species formed during a reaction.

5. In a graph of energy, I trace the way,

Mapping the journey of reactions each day.

From reactants to products, I show the flow,

In the world of mechanisms, I help you know.

Hint: I'm a diagram that shows the progress of a reaction over energy.

6. With steps and sequences, I outline the route,

Detailing reactions, I leave no doubt.

In the world of chemistry, I guide the flow,

In the dance of electrons, I help you grow.

Hint: I illustrate the step-by-step process of a chemical reaction.

7. In my mechanism, I take my stand,

A nucleophile attacks, as I planned.

Replacing a group, I change the scene,

In the world of reactions, I'm quite keen.

Hint: I'm a common type of reaction where one group replaces another.

8. In a double bond, I make my play,

An electrophile adds, and I shift the way.

Creating new bonds, I change the form,

In the realm of reactions, I weather the storm.

Hint: I typically involves the addition of an electrophile to an alkene.

9. With unpaired electrons, I dance in the night,

In a chain reaction, I take flight.

Breaking and forming bonds in a rapid race,

In the world of chemistry, I find my place.

Hint: I involve highly reactive species with unpaired electrons.

10. In a bid to simplify, I break a bond,

Removing elements, I respond.

From larger to smaller, I change my face,

In the world of reactions, I find my space.

Hint: I'm a type of reaction where elements are removed to form a double bond.

11. In my process, two come together with glee,

Adding atoms or groups, as you can see.

From double to single, I change the frame,

In the world of chemistry, I play the game.

Hint: I often occurs with unsaturated compounds.

12. With bonds that shift and atoms that sway,

I change my structure in a clever way.

Old connections break, new ones arise,

In the realm of mechanisms, I surprise.

Hint: I involves the reorganization of atoms within a molecule.

13. In my pathway, I take my time,

With a two-step dance, I'm quite sublime.

First, I form a carbocation bright,

Then a nucleophile comes to ignite.

Hint: I involves a unimolecular nucleophilic substitution.

14. In a single step, I move with grace,

Nucleophile attacks, I change my face.

With backside attack, I switch the roles,

In the world of reactions, I achieve my goals.

Hint: I's a bimolecular nucleophilic substitution.

15. With a stepwise process, I take my lead,

A carbocation forms, then I proceed.

Eliminating groups as I dance along,

In the world of chemistry, I'm where I belong.

Hint: I involves a unimolecular elimination process.

16. In one swift motion, I make my mark,

A strong base attacks, igniting the spark.

Simultaneous removal, I set the pace,

In the realm of reactions, I find my place.

Hint: I's a bimolecular elimination that occurs in one step.

17. With unpaired electrons, I start the show,

Breaking bonds apart in a chemical flow.

In chain reactions, I lead the spree,

In the world of radicals, I'm wild and free.

Hint: I involves reactive intermediates with unpaired electrons.

18. In one single step, I change and I shift,

No intermediates here, it's a molecular gift.

With bonds breaking and forming all at once,

In the realm of reactions, I'm a powerful dunce.

Hint: I involves simultaneous bond-making and bond-breaking.

19. With equal sharing, I break apart,

Bonding electrons split, that's my art.

Creating radicals, I'm on my way,

In the dance of chemistry, I come to play.

Hint: I's the process of breaking a bond to form radicals.

20. With a lopsided break, I take my turn,

Electrons go with one, it's what I learn.

I form ions as bonds come undone,

In the world of reactions, I've just begun.

Hint: I involves the breaking of a bond where one atom retains

both electrons.

21. In every reaction, I stand in the way,

The energy barrier that mustn't sway.

To overcome me, energy must be supplied,

In the journey of reactions, I'm the guide.

Hint: I'm the energy required to initiate a chemical reaction.

22. In the slowest part, I take my place,

I control the speed of the reaction race.

With every pathway, I set the tone,

In the world of kinetics, I'm well known.

Hint: I'm the step that limits the overall reaction rate.

23. In the game of reactions, I'm a helpful friend,

Lowering energy barriers, I'm here to lend.

I speed up the process, without being spent,

In the world of chemistry, I'm the event.

Hint: I increases the rate of a reaction without being consumed.

24. With an empty orbital, I seek to gain,

A pair of electrons, I'll entertain.

In the world of reactions, I'm the one,

Accepting electrons, I have my fun.

Hint: I'm a substance that accepts an electron pair.

25. With a pair of electrons, I'm ready to share,

Donating to others, I show I care.

In the dance of reactions, I take my stand,

In the world of chemistry, I lend a hand.

Hint: I'm a substance that donates an electron pair.

26. In the realm of protons, I take the lead,

Donating hydrogen, fulfilling the need.

In the world of reactions, I'm a key player,

In the game of acidity, I'm the slayer.

Hint: I'm a substance that donates a proton.

27. With a thirst for protons, I stand by,

Accepting hydrogen, reaching for the sky.

In the realm of reactions, I find my way,

In the world of acidity, I'm here to stay.

Hint: I's a substance that accepts a proton.

28. In a transfer of protons, we make our play,

An acid and a base dance the day away.

Forming water and salt, we take our stance,

In the world of chemistry, we love to dance.

Hint: I involves the transfer of protons between acids and bases.

29. In the presence of solvent, I take my cue,

Breaking apart as I bid adieu.

With solvent assisting in the reaction's race,

In the realm of chemistry, I find my place.

Hint: I's a reaction where a solvate assists in breaking a bond.

30. With water in tow, I split apart,

Breaking bonds in a chemical art.

In reactions with ease, I change the scene,

In the world of chemistry, I reign supreme.

Hint: I's a reaction where water breaks down a compound.

31. In my process, two join with a sigh,

Releasing water, as bonds unify.

Building larger from smaller, I make my stand,

In the world of chemistry, I'm perfectly planned.

Hint: I often produces water as a byproduct.

32. In a concert of bonds, I bring two together,

A diene and a dienophile, in any weather.

Forming a ring, I create something new,

In the realm of reactions, I'm quite the view.

Hint: I's a cycloaddition reaction between a diene and an alkene.

33. With aldehydes and ketones, I start my play,

Forming a β-hydroxy carbonyl along the way.

Heat it up, and I'll remove some water,

In the world of reactions, I'm a clever plotter.

Hint: I's a reaction that involves the formation of carbon-carbon
bonds.

34. With magnesium in tow, I'm ready to strike,

Forming new bonds, I do it with like.

In creating alcohols, I play my part,

In the world of synthesis, I'm quite the art.

Hint: I involves the addition of Grignard reagents to carbonyl compounds.

35. With phosphonium in hand, I start my quest,

Creating alkenes, I give it my best.

From carbonyls, I'll make a double bond,

In the realm of synthesis, I'm quite fond.

Hint: I's a method for forming alkenes from aldehydes or ketones.

36. In my reaction, I add to the game,

A nucleophile attacks, and it's never the same.

Forming bonds in a conjugated scene,

In the world of chemistry, I'm often seen.

Hint: I involves the addition of a nucleophile to an α,β-unsaturated carbonyl compound.

37. In the realm of rings, I take my stance,

Electrophiles enter, it's their chance.

Replacing a group in a dance so fine,

In the world of aromatics, I'm truly divine.

Hint: I's a reaction where an electrophile substitutes for a hydrogen on an aromatic ring.

38. In the world of bonds, I twist and I turn,

Changing my structure, there's much to learn.

With atoms shifting, I find my place,

In the chemistry game, I pick up the pace.

Hint: I's a process where a molecule transforms into another isomer.

39. With two partners, I make my play,

Joining together in a grand display.

Forming larger structures, I take my chance,

In the world of chemistry, I lead the dance.

Hint: I involves the formation of a new bond between two fragments.

40. In the process of breaking, I come to be,

Turning one into many, just wait and see.

With heat or light, I make my mark,

In the world of reactions, I leave my spark.

Hint: I's a reaction where a compound breaks down into simpler substances.

Chapter 14
Additional Terms

1. In a chain of molecules, I come alive,

Linking together, I help them thrive.

From monomers to polymers, I make my way,

In the world of materials, I'm here to stay.

Hint: I's the process of combining small units to form a large macromolecule.

2. With two types in hand, I join the fray,

Creating a blend that can play all day.

In my structure, diversity reigns,

In the world of polymers, I break the chains.

Hint: I's a polymer made from two different monomers.

3. In my connections, I make it tight,

Joining strands together, I bring the light.

Strengthening structure, I'm part of the game,

In the world of polymers, I make my name.

Hint: I's the process where polymer chains are linked together.

4. With water as my ally, I break apart,

Splitting compounds, I play my part.

In reactions I thrive, with ease and grace,

In the world of chemistry, I find my place.

Hint: I involves the chemical breakdown of a compound due to reaction with water.

5. In the study of heat, I take my stand,

Understanding energy changes, I'm quite grand.

From spontaneity to equilibrium's call,

In the world of chemistry, I cover it all.

Hint: I's the branch of physical chemistry that deals with energy changes.

6. In the race of reactions, I set the pace,

Studying rates, I keep up the chase.

Factors that influence how fast they go

In the world of reactions, I help you know

Hint: I's the study of the speed of chemical reactions.

7. In the balance of reactions, I hold my place,

Forward and reverse, a delicate grace.

At my state, concentrations stay the same,

In the world of chemistry, I'm the name of the game.

Hint: I's the state where reactants and products are formed at equal rates.

8. In response to change, I come into play,

Shifting positions, I find my way.

With stress applied, I'll adjust with care,

In the world of reactions, I'm always aware.

Hint: I states that a system at equilibrium will adjust to counteract a change.

9. In the steps of a dance, I guide the way,

Mapping the process, I'm here to stay.

With every transition and intermediate too,

In the world of chemistry, I help you view.

Hint: I describes the step-by-step sequence of a chemical reaction.

10. In the graph of reactions, I chart the scene,

From reactants to products, I keep it clean.

Showing energy changes, I plot my course,

In the world of chemistry, I'm a vital force.

Hint: I represents the energy changes that occur during a reaction.

11. In the realm of making, I come to play,

Combining components in a thoughtful way.

Creating new substances, I'm on a quest,

In the world of chemistry, I'm at my best.

Hint: I refers to the process of combining reactants to form a compound.

12. In the lab, I'm the one who separates,

Extracting the pure, while the rest waits.

From mixtures, I bring forth the desired find,

In the world of chemistry, I'm often kind.

Hint: I involves separating a compound from a mixture.

13. With care and precision, I remove the mess,

Making substances clean, I do my best.

Eliminating impurities, I shine a light,

In the world of chemistry, I make things right.

Hint: I's the process of removing contaminants from a substance.

14. In layers I travel, separating each part,

With substances moving, I play my art.

Colorful bands emerge as I make my way,

In the world of analysis, I'm here to stay.

Hint: I's a technique for separating mixtures based on their movement through a medium.

15. In the light I dance, revealing the clues,

Analyzing substances, I'm here to use.

With wavelengths and spectra, I uncover the truth,

In the world of chemistry, I'm the sleuth.

Hint: I involves the study of how light interacts with matter.

16. With ions I play, measuring mass with glee,

Identifying compounds, just wait and see.

In the world of particles, I take my stance,

In the game of chemistry, I lead the dance.

Hint: I's a technique used to determine the mass-to-charge ratio of ions.

17. In a measured dance, I add with care,

To find the endpoint, I'll always be fair.

Using solutions to calculate what's right,

In the world of analysis, I'm a guiding light.

Hint: I's a technique used to determine the concentration of a solution.

18. In the balance of moles, I find my flow,

Calculating amounts, I help you know.

From reactants to products, I set the stage,

In the world of reactions, I'm all the rage.

Hint: I involves the calculation of reactants and products in a chemical reaction.

19. In the world of solutions, I define the strength,

How much of a solute, I measure at length.

From diluted to strong, I chart the way,

In the realm of chemistry, I'm here to stay.

Hint: I's the amount of solute in a given volume of solution.

20. With water in hand, I lessen the might,

Reducing concentration, making it light.

Mixing with care, I change the scene,

In the world of solutions, I keep it clean.

Hint: I's the process of reducing the concentration of a solution.

21. In the measure of acids, I play my role,

From zero to fourteen, I define the whole.

With a scale that guides, I help you see,

In the world of chemistry, I'm key to be.

Hint: I's a measure of the acidity or basicity of a solution.

22. In the presence of change, I hold my ground,

Resisting shifts in pH, I'm always around.

With acids and bases, I find the way,

In the world of chemistry, I'm here to stay.

Hint: I's a solution that resists changes in pH upon the addition of small amounts of acid or base.

23. With energy supplied, I break apart,

Forming charged particles, it's a chemical art.

In the world of solutions, I take my place,

In the game of reactions, I quicken the pace.

Hint: I's the process by which neutral atoms gain or lose electrons to form ions.

24. In the dance of dissolving, I take my cue,

How much can be mixed? I'll show you too.

From solids to liquids, I find my space,

In the world of solutions, I set the pace.

Hint: I's the ability of a substance to dissolve in a solvent.

25. In the world of mixtures, I'm the fluid base,

Dissolving the solute, I'm in the race.

From water to alcohol, I find my role,

In the game of solutions, I'm on a roll.

Hint: I's the substance that dissolves a solute to form a solution.

26. In the lab, I'm a helper, a chemical friend,

Reacting with others, on me you depend.

In synthesis or tests, I play my part,

In the world of chemistry, I'm a work of art.

Hint: I's a substance used in a chemical reaction to detect, measure, or produce other substances.

27. In the end, I'm the result of the game,

From reactants to me, I take the fame.

What's formed in the reaction, I proudly show,

In the world of chemistry, I'm the star of the show.

Hint: I's the substance formed as a result of a chemical reaction.

28. In the shadows of reactions, I often appear,

Not the main goal, but still, I'm here.

A product of process, I tag along,

In the world of chemistry, I'm part of the song.

Hint: I's a secondary product formed in a chemical reaction.

29. In the measure of products, I take my stand,

How much is produced? I help you understand.

From theoretical to actual, I show the way,

In the world of reactions, I'm here to stay.

Hint: I's the amount of product obtained from a reaction compared to the theoretical maximum.

30. In the realm of reactions, I make my choice,

Favoring one pathway, I raise my voice.

With precision in mind, I guide the way,

In the world of chemistry, I'm here to stay.

Hint: I refers to the preference of a reaction to form one product over others.

31. In the dance of reactions, I choose my mate,

Favoring pathways, I decide their fate.

With precision I work, not random at all,

In the world of chemistry, I heed the call.

Hint: I refers to the preference of an enzyme or reagent for a particular substrate.

32. I bring new traits to the chemical scene,

Modifying structures, I'm the unseen queen.

Adding groups to the core, I take my chance,

In the world of synthesis, I lead the dance.

Hint: I involves the introduction of functional groups into a compound.

33. In the twist of a bond, I change my face,

From one configuration to another place.

With chiral centers, I make my play,

In the realm of molecules, I'm here to stay.

Hint: I refers to the inversion of stereochemistry at a chiral center.

34. I measure success in the chemical art,

How much product forms plays a key part.

From theoretical dreams to what's real,

In the world of reactions, I help you feel.

Hint: I's the amount of product obtained compared to the expected amount.

35. In the efficiency game, I take my score,

How many atoms are used? I show you more.

In reactions I guide, keeping waste at bay,

In the world of synthesis, I lead the way.

Hint: I refers to the measure of how well atoms are utilized in a reaction.

36. On the path of creation, I chart the way,
Mapping the steps for what you'll convey.
From start to finish, I guide the plan,
In the world of chemistry, I'm your best fan.
Hint: I outlines the series of reactions needed to produce a target compound.
37. In the mix of reactions, I take a detour,
Producing a product, but not the main score.
An unintended path, I show my face,
In the world of chemistry, I find my place.
Hint: I's an unintended reaction that occurs alongside the desired reaction.

38. With a keen eye, I observe the flow,
Unraveling pathways that reactions show.
Step by step, I seek to explain,
In the realm of chemistry, I'm the detective's gain.
Hint: I involves the investigation of how a reaction occurs at the molecular level.

39. In the lab, I lay the ground to explore,
Structuring tests, so results will pour.

With careful controls, I frame the scene,

In the world of science, I keep it clean.

Hint: I's the process of planning a controlled experiment to test a hypothesis.

40. In the realm of particles, I show my might,

Exploring the behavior of electrons in flight.

With wave functions and states, I take my stand,

In the world of chemistry, I'm truly grand.

Hint: I's the study of how quantum mechanics applies to chemical systems.

PHYSICAL

CHEMISTRY

Chapter 15

Thermodynamics

1. I measure the warmth, both high and low,

In Celsius or Kelvin, my values flow.

What am I?

Hint: I help you know if it's hot or cold outside.

2. In transfers I move, from one to the next,

A flow of energy, I'm often perplexed.

What am I?

Hint: I can make things warm, or even quite hot.

3. I'm done when a force moves through a distance,

In energy's dance, I play my existence.

What am I?

Hint: I'm involved in lifting and pushing, too.

4. Within every substance, I quietly dwell,

A sum of all motions, can you hear me tell?

What am I?

Hint: I change with heat and work, can you see?

5. I'm a measure of heat, at constant pressure,

A sum of internal energy, with a treasure.

What am I?

Hint: My symbol is H, in reactions I play.

6. In chaos I thrive, disorder my game,

A measure of randomness, that's my claim to fame.

What am I?

Hint: I always increase, or so they say.

7. In spontaneity, I hold the key,

Predicting the fate of reactions with glee.

What am I?

Hint: If I'm negative, the process is free.

8. At constant volume, my presence is key,

I help you find balance, just wait and see.

What am I?

Hint: My symbol is A, I'm linked with entropy.

9. Energy can't vanish, it's all conserved,

In transformations, my truth is served.

What am I?

Hint: It's all about energy, you see, my friend.

10. In every process, I dictate the flow,

From order to disorder, you'll see it grow.

What am I?

Hint: I explain why heat moves from hot to cold.

11. At absolute zero, my rules come to play,

Perfect order is what I say.

What am I?

Hint: The temperature is lowest, and entropy is at bay.

12. In a perfect world, I can return,

My path retraced, there's much to learn.

What am I?

Hint: I'm a process that can go both ways.

13. Once I'm done, I cannot go back,

A one-way street is my track.

What am I?

Hint: Once I take my destined path, no force can stop me.

14. I exchange energy, but not mass,

Within my boundaries, all things amass.

What am I?

Hint: I can't let anything in or out, just so.

15. I take in and give out, a flow that's free,

Mass and energy dance with glee.

What am I?

Hint: Think of a pot boiling on the stove.

16. I'm cut off from the world, sealed tight,

Neither mass nor energy can take flight.

What am I?

Hint: Think of a thermos that keeps things the same.

17. In my world, the path doesn't matter,

Only the start and end, I'll never scatter.

What am I?

Hint: Think of energy values that don't depend on how you got there.

18. I depend on the journey, not just the end,

The way you arrive is where I extend.

What am I?

Hint: Think of work and heat, they follow their own way.

19. No heat enters here, the system is sealed,

All energy shifts, in work it's revealed.

What am I?

Hint: Imagine a quick compression, where heat can't escape.

20. At constant temperature, I play my part,

Exchanging heat, that's just the start.

What am I?

Hint: Picture a slow melting ice cube in the sun.

21. At constant pressure, I make my move,

With heat exchange, my energy will improve.

What am I?

Hint: Think of a boiling pot that keeps pressure steady.

22. Volume unchanged, I'm stuck in my space,

Heat added or removed, I keep my place.

What am I?

Hint: Imagine a sealed container that can't expand.

23. Solid, liquid, gas—my states can shift,

But at balance, I give a special gift.

What am I?

Hint: Think of water at its boiling point, steady and clear.

24. I measure the potential of a substance's fate,

In solutions and reactions, I'll help you relate.

What am I?

Hint: I guide where things will go in a mix.

25. I speak of components, each in their role,

In mixtures I'm vital, that's how I console.

What am I?

Hint: Think of how each ingredient affects the whole.

26. I describe the phase change, with pressure and heat,

A vital connection, so neat and complete.

What am I?

Hint: I relate vapor pressure to temperature, so true.

27. In the realm of engines, I'm the best you'll see,

With efficiency rules, I set the degree.

What am I?

Hint: I'm a theoretical cycle that can't be beat.

28. I measure how much heat it takes to raise,

The temperature of a substance in many ways.

What am I?

Hint: Think of boiling water and how long it takes.

29. Per mole I'm measured, a specific heat,

For substances known, I can't be beat.

What am I?

Hint: I'm useful for knowing how much heat to add.

30. In a defined condition, I hold my ground,

Under standard conditions, my values are found.

What am I?

Hint: Think of elements at 1 atm and 25 degrees.

31. When one mole is formed from its elements pure,

I give the heat change, that's for sure.

What am I?

Hint: My values help in reaction calculations.

32. I measure disorder at standard conditions,

A crucial part of thermodynamic decisions.

What am I?

Hint: I always increase as systems become more random.

33. A map of states where pressure and heat meet,

I guide through transitions, a visual treat.

What am I?

Hint: Think of solid, liquid, gas boundaries drawn out.

34. In a closed space, I exert my might,

A pressure from vapor, day or night.

What am I?

Hint: I increase with temperature, can you see?

35. At this juncture, gas and liquid collide,

No distinction remains, they merge and abide.

What am I?

Hint: Beyond me, properties are hard to divide.

36. I'm the heat exchanged with no temperature rise,

In phase changes I'm crucial, a real surprise.

What am I?

Hint: Think of boiling water where temperature stays still.

37. In reactions I'm measured, the change you can see,

The energy shift as bonds break or agree.

What am I?

Hint: I can be positive or negative, depending on the reaction's

spree.

38. In my presence, the path is ignored,

Total heat change, no matter the hoard.

What am I?

Hint: Think of breaking down a journey into steps.

39. As gases expand, I show a twist,

A cooling or heating, you can't resist.

What am I?

Hint: Think of a gas escaping from a pressurized container.

40. When a liquid turns to vapor, I come into play,

The heat required, I measure each day.

What am I?

Hint: My value is high for water, a crucial phase change.

41. I'm the heat needed for melting, that's true,

Solid to liquid, I'm part of the view.

What am I?

Hint: Ice requires me to become water, you see.

42. In the realm of reactions, I map the way,

From reactants to products, I show the sway.

What am I?

Hint: Think of a graph that shows how energy shifts.

43. At my peak, reactions are poised to break,

An unstable point, it's the path we take.

What am I?

Hint: I exist momentarily, before the products awake.

44. The barrier to overcome, I stand in the way,

The energy needed for reactions to play.

What am I?

Hint: Without me, reactions might take forever to sway.

45. In the realm of reactions, I dictate the ease,

Whether a process will happen, I aim to please.

What am I?

Hint: I'm tied to Gibbs energy, helping you see.
46. In perfect balance, I hold my reign,
Energy flows equally, without any strain.

What am I?

Hint: Imagine a still pond, where everything is plain.

Chapter 16
Chemical Kinetics

1. I measure how fast molecules race,

The speed at which they set the pace.

Can you name me, this key metric,

In the study of all things kinetic?

Hint: How fast does a change in concentration take place?

2. A formula that tells how fast,

The outcome you can forecast.

Look at concentrations with care,

Then use me to compare.

Hint: It involves a relation between rate and concentration.

3. I'm the factor that multiplies all,

Changing with heat, sometimes small.

No units fixed, yet always there,

I help reactions move, or stay where.

Hint: I'm specific to every reaction at a certain temperature.

4. First, second, or even none,

My value tells how it's done.

From concentrations, I take my cue,

What reaction order could this be true?

Hint: This determines the overall power in the rate law.

5. In this case, the rate won't change,

No matter how much you rearrange.

Add more reactant, none will care,

This unique order is quite rare.

Hint: The concentration does not affect the rate.

6. One reactant controls the race,

Its concentration sets the pace.

A simple power tells it all,

Just one step and I won't stall.

Hint: The exponent is one in the rate law.

7. When two collide to make a mark,

The rate goes high and hits its arc.

A square or pair makes me unique,

Which order does this reaction seek?

Hint: It could depend on one reactant squared or two reactants.

8. Half of what you start with goes,

The time it takes? Who really knows?

But I measure this decline,

And help reactions stay in line.

Hint: How long does it take for half of the reactants to disappear?

9. I'm the wall reactants must pass,

Without me, they'll run out of gas.

Add some heat to cross the line,

And let the reaction work just fine.

Hint: Think of the energy needed to get started.

10. With heat and constants, I predict,

How fast reactions might be kicked.

From energy hills to molecule might,

I show how temp changes the fight.

Hint: Temperature and activation energy play key roles here.

11. I speed things up without a change,

I stay the same through the range.

A helper in the reaction flow,

Who am I? Can you know?

Hint: I lower the activation energy without getting consumed.

12. Steps I hide, yet show in clues,

Paths reactions always choose.

From start to end, I map the way,

Find my name, what do you say?

Hint: I describe the sequence of elementary steps.

13. At the top, I stand so tall,

Neither here nor there at all.

I'm fleeting, gone in just a flash,

But without me, reactions crash.

Hint: I represent the peak energy point of the reaction.

14. I'm a single, simple act,

Just one step, that's a fact.

Together we form the bigger play,

Name me, then be on your way.

Hint: The simplest step in a mechanism.

15. I appear for a short-lived time,

Not in the end, I'm a fleeting rhyme.

I vanish before the final show,

Who am I, do you know?

Hint: I exist only between the steps of a mechanism.

16. When molecules meet and greet,

Energy and angle must be neat.

If they strike just right, sparks fly,

What's my name, do you spy?

Hint: I explain how molecules must collide to react.

17. My name tells how many must meet,

For a step to be complete.

One, two, or three at most,

Find me fast before I'm lost.

Hint: I refer to the number of molecules involved in an elementary step.

18. I keep things constant in between,

Though intermediates are rarely seen.

A simplifying trick for rates,

What's my name that calculates?

Hint: It assumes that intermediate concentration stays constant.

19. Slow and steady wins the race,

But sometimes I set the pace.

The slowest step in a reaction's tale,

What is this? Don't let it fail.

Hint: The slowest step in the mechanism.

20. One step sparks another flame,

In a series, I take my name.

A repeating loop until the end,

Can you guess what I intend?

Hint: Each step regenerates a species that continues the process.

21. I'm a catalyst, but quite refined,

I make reactions faster and aligned.

With substrates bound in a perfect fit,

What's my field? Can you admit?

Hint: Biological molecules act as catalysts here.

22. With rates and enzymes, I describe,

How reactions with proteins thrive.

Substrate concentrations in the mix,

What's this formula's hidden fix?

Hint: It explains enzyme-catalyzed reactions.

23. In the same phase, I do reside,

Reactants and I side by side.

Together, we make reactions zoom,

Who am I in the reaction room?

Hint: The catalyst and reactants are in the same phase.

24. On a surface, I stand tall,

Reactants visit me, that's all.

Different phases, that's my game,

Now try to guess my name.

Hint: The catalyst and reactants are in different phases.

25. A graph that shows the ups and downs,

Through valleys deep and hilltop crowns.

It maps the energy from start to end,

What's my name? Now comprehend.

Hint: I chart the path of energy throughout the reaction.

26. A picture of peaks and valleys deep,

Shows the energy each must keep.

The higher I go, the harder to fight,

What's my graph? I show the height.

Hint: I plot energy versus reaction progress.

27. Before the climb, I set the stage,

A number to help in every gauge.

With activation energy in play,

Who am I in this array?

Hint: I'm part of the Arrhenius equation, determining how frequently molecules collide.

28. More of me, and faster I go,

Add less, and I'm rather slow.

I'm a key to reaction speed,

Who am I in this deed?

Hint: Reactants' amount directly influences the rate.

29. Heat me up, and I increase,

Lower the heat, and I release.

I speed reactions up with heat,

What's my name? Now be discreet.

Hint: Raising this can often speed up reactions.

30. I take the form and shape of time,

Calculating changes line by line.

For different orders, I give a clue,

Which law am I, can you view?

Hint: I relate time to concentration in various reactions.

31. More than one step, that's my style,

With intermediates and steps compiled.

I'm no simple one-step show,

What's my name? You should know.

Hint: Reactions with multiple elementary steps.

32. Two or more paths from one same start,

Leading to products far apart.

At the same time, both can flow,

What's my type? Do you know?

Hint: Reactants split into multiple different products.

33. One reaction follows on the other's tail,

In a sequence, I prevail.

Products turn to reactants new,

What's my name? Can you view?

Hint: Reactions that occur step-by-step in sequence.

34. I move forward, then I reverse,

Reactants and products freely traverse.

In balance, I like to stay,

What's my name? What do you say?

Hint: Reactions that can go both ways.

35. When forward and backward match in pace,

Neither side gains or loses its place.

A balance in reactions set,

Can you name this state we met?

Hint: The rate of forward and reverse reactions are equal.

36. I'm not still, though balanced fine,

Reactions occur all the time.

Forward and backward in harmony dance,

What's this dynamic reaction stance?

Hint: Both reactions continue, but the concentrations stay constant.

37. I slow things down or block the way,

I make sure reactions delay.

Without me, they'd run too fast,

What am I, holding them last?

Hint: I decrease the rate of a reaction.

38. I speed myself up as I go,

The more I make, the faster I flow.

A self-helper in the mix,

Who am I, can you fix?

Hint: The product of a reaction catalyzes the reaction itself.

39. I don't settle for just one state,

Changing forms at every rate.

Back and forth, in time I swing,

What's my name, this rhythmic thing?

Hint: Reactions that cycle between different states over time.

40. Before equilibrium, I hold the key,

A snapshot of what's come to be.

Compare me to Keq to know,

Which direction reactions will go.

Hint: I compare concentrations at any given time to the equilibrium constant.

41. One spark makes many more,

In rapid splits, I explore.

A chain reaction with many fates,

What's my name? It correlates.

Hint: Each step leads to multiple reactive intermediates.

42. I tell of energy, heat, and flow,

But in the realm of rates I also show.

Heat and spontaneity decide the fate,

Who am I in this debate?

Hint: I bridge the concepts of heat and reaction rates.

43. I show how rates and changes relate,

For different orders, I calculate.

Instantaneous rates you can now find,

What's my law? Keep me in mind.

Hint: I express the rate of reaction as a function of concentration.

44. Collisions help me on my way,

But not all make it through the day.

A two-step path to explain the slow,

What's my theory? Do you know?

Hint: I explain unimolecular reactions through a two-step process.

45. One molecule stands on its own,

No partner needed to be shown.

I break apart or rearrange,

What reaction am I in exchange?

Hint: Only one molecule reacts to form the product.

46. Two come together in a flash,

Collide and create, then they dash.

Their meeting leads to something new,

What type of reaction do I cue?

Hint: Two reactant molecules are involved.

47. The faster we meet, the faster we react,

In the solution, it's an established fact.

I depend on movement, fast and free,

What's my reaction type? Can you see?

Hint: The rate depends on how quickly molecules can diffuse through a medium.

48. Experiment and math combine,

To find out how reactants align.

What's the name of this careful quest,

To determine the rate's real test?

Hint: I find how concentration affects rate by experimentation.

49. To simplify and understand,

I hold all but one reactant in hand.

Changing just one, I measure the speed,

What's this method for rate law need?

Hint: I isolate one reactant's effect on the rate by holding others constant.

50. I show how molecules shift and change,

Each step mapped within my range.

I tell the story of how things go,

What's my path? What do you know?

Hint: I describe the detailed steps of how a reaction proceeds.

Chapter 17
Chemical Equilibrium

1.I tell you how far reactions go,

Whether products dominate the show.

From reactants to products, I dictate the game,

What constant holds my famous name?

Hint: I compare the concentrations of products to reactants at equilibrium.

2. I look at things at any stage,

Am I ahead, or behind the gauge?

Compare me to K, what will I do,

Shift forward or backward, which way do I cue?

Hint: I determine whether a reaction needs to move forward or backward to reach equilibrium.

3. Disturb me, and I'll fight back,

Push me here, I'll shift the track.

Change the pressure, volume, or heat,

I'll adjust and never miss a beat.

Hint: I predict how equilibrium adjusts when conditions change.

4. Though still, I am always in motion,

Reactants and products in fluid devotion.

Forward and reverse, they move just right,

Can you guess this constant fight?

Hint: Reactions continue, but the concentrations remain unchanged.

5. In the same phase, I always dwell,

Reactants and products where stories swell.

One phase for all is how I play,

What's my type? Can you say?

Hint: Reactants and products exist in the same phase.

6. In more than one phase, I take my stand,

With solids, liquids, and gases at hand.

Different states, yet balanced still,

What am I? Can you tell?

Hint: Reactants and products exist in different phases.

7. Moving from left to right,

I lead the reactants into the light.

Products form as I proceed,

Which reaction am I? Indeed!

Hint: I describe the process where reactants turn into products.

8. From products, I make my way back,

Turning them into reactants in my track.

I go against the forward tide,

What reaction am I? What's my guide?

Hint: I describe the process of products converting back to reactants.

9. Concentrations are my guide,

For equilibrium, I provide.

A ratio for reactions great or small,

What law governs them all?

Hint: I express the relationship between concentrations of reactants and products at equilibrium.

10. Each gas has its share to claim,

In mixtures, we're never the same.

I sum up to total pressure's call,

What am I for one, among them all?

Hint: I represent the pressure exerted by an individual gas in a mixture.

11. I measure how much is in a space,

More of me can shift the race.

Reactants or products, I don't stay still,

What do I measure, what's the drill?

Hint: I quantify the amount of a substance in a given volume.

12. Add some heat, or cool things down,

I make reactions wear a frown.

Shift equilibrium this way or that,

What do I cause with heat in chat?

Hint: Raising or lowering me changes the balance of reactions.

13. Squeeze me hard or let me go,

Gases shift as pressures grow.

Equilibrium responds to my might,

What effect am I in this fight?

Hint: Increasing or decreasing volume affects gaseous reactions at equilibrium.

14. When I expand or shrink a space,

Equilibrium shifts to find its place.

For gases alone, I hold the key,

What change am I? Can you see?

Hint: Changes in this for gases affect equilibrium.

15. I make things fast without a change,

But equilibrium? I don't rearrange.

I speed reactions up or slow them down,

Who am I in this crown?

Hint: I affect reaction speed but not the final equilibrium position.

16. I tell you if reactions will flow,

Spontaneous or not, I let you know.

Linked to equilibrium with a clever hand,

What energy do I command?

Hint: I determine the spontaneity of a reaction and its relation to equilibrium.

17. Energy flows out as I proceed,

Products form with a burst of speed.

I happen on my own, no need to shove,

What type of reaction do you love?

Hint: Reactions where energy is released and ΔG is negative.

18. I need a push, I don't go free,

Energy must be fed to me.

Reactants sit until energy flows,

What's my reaction? Do you know?

Hint: Reactions where energy is absorbed and ΔG is positive.

19. At standard conditions, I take my place,

A constant that helps reactions trace.

A special form of K you see,

Who am I in chemistry?

Hint: I apply at standard pressure and temperature.

20. Where does the balance lie for me,

More reactants or products do we see?

This point defines where I reside,

What's my name? Do you decide?

Hint: I describe the ratio of products to reactants at equilibrium.

21. For solids dissolving in a fluid bath,

I define the solubility path.

With a special constant all my own,

What's my name, can it be known?

Hint: I describe the extent to which a solid dissolves in water.

22. I shift solubility when I'm around,

Adding ions makes the product drown.

Precipitates form when I appear,

What's my effect? Now make it clear.

Hint: I reduce solubility by adding an ion already present in solution.

23. I keep pH steady, no matter the fight,

Adding acids or bases, I stay right.

A mixture of weak, that's my name,

What solution holds steady in the game?

Hint: I resist changes in pH when small amounts of acids or bases are added.

24. I break apart when water's near,

Into ions I disappear.

Charged particles form as I decay,

What's this process in my display?

Hint: I describe the process of forming ions from neutral molecules.

25. I show how much is left intact,

Or how much ionizes, that's a fact.

A fraction of the whole you'll see,

What am I in chemistry?

Hint: I represent the fraction of molecules that dissociate in solution.

26. Between solid, liquid, and gas I play,

At a certain point, I stay.

In balance between phases, I lie,

What's my name? Come, give a try.

Hint: I describe the state where different phases coexist in equilibrium.

27. When something changes in the mix,

I respond with a clever fix.

I move to restore what I had before,

What am I when equilibrium restores?

Hint: I occur when a disturbance causes a change in equilibrium position.

28. When forward and backward match in speed,

Neither gains, nor one will lead.

Rates equal in this balanced fight,

What is this, can you get it right?

Hint: I describe when the rate of forward and reverse reactions
are equal.

Chapter 18

Electrochemistry

1. Two processes, one gives, one takes,

Together, they form a balanced shake.

Electrons move in a lively game,

What's the reaction? Can you name?

Hint: The combination of oxidation and reduction in one reaction.

2. I lose electrons, energy I make,

In this process, it's my stake.

Oxygen may join, or just electrons go,

What's my name? Do you know?

Hint: The process of losing electrons.

3. I gain electrons, energy stored,

My role in redox is never ignored.

Less oxygen or more hydrogen for me,

What's this process called, can you see?

Hint: The process of gaining electrons.

4. I help others to lose and burn,

But in the process, I take my turn.

I get reduced, gaining what's lost,

What's my role? Guess the cost.

Hint: I make oxidation happen by gaining electrons.

5. I give electrons with open hand,

Causing reduction, as I had planned.

In return, I'm oxidized, you see,

What am I, can you tell me?

Hint: I cause reduction by losing electrons.

6. A setup where electrons flow,

Chemical energy to electric show.

Two half-cells that work in unison well,

What type of device? Can you tell?

Hint: I convert chemical energy into electrical energy.

7. Spontaneous reactions power my course,

Electric current from a chemical source.

Anode to cathode, electrons flee,

What's my name? Do you agree?

Hint: I generate electrical energy from a spontaneous reaction.

8. With energy, I force the flow,

Non-spontaneous reactions start to grow.

Electric current makes things change,

What cell am I within this range?

Hint: I use electrical energy to drive a non-spontaneous reaction.

9. I'm where reactions meet and greet,

On my surface, the changes beat.

Cathode or anode, that's my role,

What am I in this electro-goal?

Hint: I'm a conductor where oxidation or reduction occurs.

10. At my surface, electrons leave,

I'm where oxidation you perceive.

In electrolytic or galvanic ways,

What's my name in these arrays?

Hint: I'm the electrode where oxidation occurs.

11. Electrons come to rest at me,

Reduction happens here, you'll see.

I complete the circuit's electric sway,

What electrode am I today?

Hint: I'm the electrode where reduction occurs.

12. I keep the charge from going astray,

I balance ions, and pave the way.

Without me, the circuit's not complete,

What am I, holding the seat?

Hint: I maintain charge balance between two half-cells.

13. I drive electrons through the wire,

The force behind their electric fire.

Measured in volts, I show my might,

What's my name? Am I right?

Hint: I measure the voltage difference between two electrodes.

14. At standard conditions, I shine bright,

I tell if reactions go left or right.

Compare me to hydrogen, see my worth,

What's my potential on this Earth?

Hint: I measure the tendency of an electrode to gain or lose electrons under standard conditions.

15. When conditions change, I step in,

With concentration, I begin.

I adjust potential as needed here,

What's this equation I hold dear?

Hint: I calculate the cell potential under non-standard conditions.

16. With electric current, I break bonds,

Forcing reactions with strong responds.

Water to hydrogen, metals to pure,

What's the name of this procedure?

Hint: I use electricity to drive a chemical reaction.

17. Electrons flow, matter's my guide,

Quantities relate to charges applied.

I predict how much will change,

What laws are these within my range?

Hint: I relate the amount of substance transformed during electrolysis to the amount of electricity passed.

18. I coat objects, shiny and new,

With electric currents, metals accrue.

Gold, silver, copper, and more,

What's this process you're looking for?

Hint: I use electrolysis to coat one metal onto another surface.

19. Through a solution, ions flow,

How well they move, I let you know.

Charged particles lead the way,

What's my name? Can you say?

Hint: I measure how well a solution conducts electricity.

20. Concentration affects how well I run,

When divided by moles, the work is done.

I show the flow for every mole,

What am I in this role?

Hint: I measure conductivity per mole of electrolyte in solution.

21. Two half-cells with the same kind,

But concentrations differ in mind.

Electrons flow to equalize,

What cell am I? Here's a surprise!

Hint: I generate a potential difference from differing concentrations of the same electrolyte.

22. I power things with simple grace,

Fuel flows in, and electrons race.

From hydrogen or methane, I get my drive,

What's my name? I keep things alive.

Hint: I generate electricity through the reaction of fuel with an oxidant.

23. Unwanted reactions with air or sea,

Rust and tarnish, you see me.

I slowly destroy what was once strong,

What's this process? Am I wrong?

Hint: I involve the oxidation of metals over time.

24. Oxidation here, reduction there,

Together, we're a balanced pair.

Each side tells its own tale,

What reactions prevail?

Hint: I describe the two parts of a redox reaction happening in separate compartments.

25. Stored energy inside my core,

I power devices more and more.

Galvanic cells within me stay,

What am I in the modern day?

Hint: I store chemical energy and convert it to electrical energy when needed.

26. I work once, then I'm done,

Use me up, I'm not much fun.

A one-time battery, that's my game,

What's my name? Guess my claim.

Hint: I'm a non-rechargeable battery.

27. I store my energy, then let it out,

But charge me up, I turn about.

Over and over, I can perform,

What type of battery is my norm?

Hint: I can be recharged after use.

28. Extra voltage I require,

Above what's expected in desire.

I slow reactions more than they should,

What am I? Am I understood?

Hint: I'm the extra voltage needed to drive a reaction beyond its theoretical potential.

29. Sensitive to just one kind,

Of ions, I can find.

Measuring concentration's my task,

What electrode am I? You might ask.
Hint: I detect the concentration of a specific ion in a solution.

214

ANALYTICAL CHEMISTRY

Chapter 19
Qualitative Analysis

1. Positive charge, I hold with pride,

Testing solutions, I don't hide.

Divided into groups, I stand tall,

What's the name of this analysis call?

Hint: This involves identifying positively charged ions.

2. Negatively charged, I roam free,

To find me, you'll test with glee.

Halides, sulfates, and more to see,

What's the analysis called for me?

Hint: This involves identifying negatively charged ions.

3. In water, two solutions mix,

A solid forms, like magic tricks.

The product falls from the solution's grace,

What reaction occurs in this case?

Hint: A solid forms when two soluble compounds react.

4. A measure of what can dissolve or stay,

In equilibrium, I hold sway.

If too much enters, I precipitate,

What constant controls this fate?

Hint: I determine the maximum concentration of ions in a saturated solution.

5. With heat, I change my hue,

My color reveals what's inside you.

Sodium is yellow, copper is blue,

What test is this? Have a clue?

Hint: I reveal metal ions through the color of flame.

6. After clues and signals, I reveal,

The final truth, the real deal.

I confirm what's hidden within the mix,

What test does this refer to, in the fix?

Hint: This is done to verify the presence of a specific ion.

7. A drop or two, that's all I need,

To tell you what's inside, indeed.

On paper or plate, I make my stand,

What quick test am I, so grand?

Hint: A small-scale test using a reagent drop to detect ions.

8. I form a bond, not just with one,

But many atoms in the fun.

Ligands join, a structure to see,

What's this reaction that involves me?

Hint: I involve multiple bonds forming between metal ions and ligands.

9. I help reactions come to light,

With my touch, the change is bright.

A test I make, with you I blend,

What's my name, my chemical friend?

Hint: I'm a chemical added to provoke a reaction.

10. To divide the parts, I come in play,

To isolate each in a clear array.

Filtration, distillation—methods all,

What techniques are these you call?

Hint: I help separate mixtures into components.

11. Through a porous screen, I pass,

To separate solid from liquid mass.

One stays behind, the other goes,

What process is this, do you know?

Hint: I separate solids from liquids by passing through a filter.

12. I spin so fast to make things clear,

Heavy parts settle, light parts steer.

What's this method of separation true,

That I perform, what do I do?

Hint: I use high-speed spinning to separate particles based on density.

13. Without much fuss, I pour with ease,
To leave the sediment, if you please.
What's this simple method I employ,
To separate mixtures without much joy?
Hint: I involve pouring off the liquid, leaving the sediment behind.

14. When two opposites meet, they react,
Protons move, as a matter of fact.
Salt and water may form as end,
What's the name of this reaction, my friend?
Hint: I involve the transfer of protons between substances.

15. One ion leaves, the others stay,
To form a solid in a clever way.
By choosing the right reagent on hand,
What technique is this? Do you understand?
Hint: I precipitate only one ion from a mixture using specific reagents.

16. I target groups with one big strike,
To find the ions I specifically like.
Split by charges, in steps I go,

What's the name of the reagents I show?

Hint: I'm used to identify ion groups in systematic qualitative analysis.

17. Each ion has a signal or clue,

Tests reveal what's inside you.

From flame to precipitation true,

What process am I leading you through?

Hint: I involve different tests to find out which ions are present.

18. With no numbers, but signs I seek,

Colors, precipitates—the clues unique.

Inorganic ions I aim to find,

What's this process that comes to mind?

Hint: I identify inorganic substances based on their chemical reactions.

19. On paper or column, I make my way,

Separating colors and compounds in play.

The components travel at different speeds,

What's the method that fulfills these needs?

Hint: I separate mixtures based on differential movement in a medium.

20. With a pop or flame, I reveal,

What invisible presence you feel.

Gas bubbles rise, but which is near,

What kind of test makes it clear?

Hint: I detect gases using simple observable reactions (like splints and limewater).

Chapter 20
Quantitative Analysis

1. I mix and measure, drop by drop,

To find the balance where reactions stop.

With a standard solution in hand,

What technique am I in this grand?

Hint: A method to determine concentration by neutralization.

2. A known concentration, precise and true,

I help in titrations and tests anew.

Carefully prepared, I guide your way,

What's my name in this analytical play?

Hint: I contain a known amount of solute in a given volume.

3. I'm a measure of concentration, you see,

Moles per liter is how I decree.

In solutions, I help you relate,

What's this term that defines my state?

Hint: I express concentration in moles of solute per liter of solution.

4. With equivalents I define my scope,

In acids and bases, I give you hope.

Molarity adjusted for the reaction's game,

What's this measure? Can you name?

Hint: I measure concentration in equivalents per liter.

5. Pure and stable, I reign supreme,

For titrations, I fulfill the dream.

My weight gives exactness to your quest,

What's my title? I'm the best!

Hint: I'm a substance with a known high purity used for calibration.

6. I may not be pure, but useful still,

Calibrated against a primary skill.

I help in analyses where accuracy's sought,

What am I? Can you think a thought?

Hint: I'm a solution prepared from a primary standard for use in titrations.

7. A color change or signal bright,

Signals the finish, the end of the fight.

I tell you when the reaction is done,

What's my name in this analytical run?

Hint: The point in titration where a change indicates completion.

8. At this stage, moles are equal in sway,

Acid meets base in a balanced way.

No excess remains, all has combined,

What's this point in the analytical mind?

Hint: The point in titration where the amounts of reactants are stoichiometrically equivalent.

9. With color shifts, I'm on the scene,

Helping chemists know what's routine.

I change hue at the right pH,

What am I, in this analytical reach?

Hint: A substance that changes color at a specific pH or concentration.

10. By weighing my solids, I find my worth,

Precipitation brings new life to birth.

With mass measurements, I can tell,

What's this analysis? Can you spell?

Hint: A method involving the measurement of mass to determine the quantity of an analyte.

11. I measure volume with precision so fine,

In titrations and dilutions, I shine.

What's this technique that counts every drop,

To find the concentration, I can't stop?

Hint: A method of analysis based on measuring volumes.

12. With data points plotted, I guide your way,

Relating concentration to signals at play.

In graphs, I show how to interpret the line,

What's this curve that helps you define?

Hint: A graphical representation of the relationship between concentration and analytical response.

13. The lowest amount I can reveal,

Below which signals are hard to feel.

In your analysis, I set the tone,

What's this limit? Can you make it known?

Hint: The smallest concentration of an analyte that can be reliably detected.

14. I'm higher than LOD, more clear and bright,

The lowest amount you can measure just right.

With confidence, I provide the read,

What's this limit? Can you take heed?

Hint: The lowest concentration of an analyte that can be quantitatively determined.

15. In my presence, impurities lay low,

I tell you how pure your substance may show.

A percentage measure, clear and precise,

What's my name, in this analytical slice?

Hint: The ratio of the mass of the pure substance to the total mass, expressed as a percentage.

16. With precision I weigh, to a thousandth degree,

Ensuring your results are accurate, you see.

What's my name, in the lab's dance,

When measuring masses, I enhance?

Hint: A scale that measures mass with high precision.

17. With light, I measure, absorbance in sight,

Analyzing samples, I make things bright.

At specific wavelengths, I work my charm,

What's this technique, can you disarm?

Hint: I measure the amount of light absorbed by a solution at specific wavelengths.

18. With concentration and path length combined,

I relate absorbance to what you find.

A linear relationship, clear and neat,

What's the law that makes this complete?

Hint: I describe how absorbance is proportional to concentration.

19. In measurements, I'm the repeatable friend,

Consistent results, I help you extend.

What's my name when you measure with care,

Ensuring results are always fair?

Hint: I refer to how close multiple measurements are to each other.

20. In the realm of truth, I hold my ground,

How close to the target I can be found.

What's my name in the quest for right,

To measure with care, shining bright?

Hint: I indicate how close a measured value is to the true value.

Chapter 21
Spectroscopy

1. In a world of light, I play my part,

How much is taken? That's my art.

With greater values, I make things clear,

What's my name when absorption's near?

Hint: I measure how much light is absorbed by a sample.

2. Light passes through, I keep track,

The percentage that makes its way back.

When little gets through, I stand low,

What's my name in this spectral show?

Hint: I express the fraction of light transmitted through a sample.

3. In a sea of colors, I measure the flow,

From crest to crest, my distance will show.

In nanometers, my value shines,

What's my name among the lines?

Hint: I'm the distance between two consecutive peaks of a wave.

4. I analyze light, near and far,

With prisms or gratings, I help you spar.

In labs, I'm your guiding star,

What am I in this analytical bazaar?

Hint: I'm an instrument that measures the intensity of light at various wavelengths.

5. A rainbow of colors, I show in array,
From UV to IR, I reveal the way.
With each line, a story to tell,
What's this display? Can you spell?
Hint: I'm a graphical representation of light intensity versus wavelength or frequency.

6. I relate concentration to the light's fate,
How much is absorbed at each rate.
With units to guide you, I stand proud,
What's my title in this spectral crowd?
Hint: I'm a constant that indicates how strongly a substance absorbs light at a given wavelength.

7. With points plotted high and low,
I show the relationship you need to know.
From concentration to signal, I'm the guide,
What's this curve that helps you decide?
Hint: A graph relating the known concentrations of a solution to its measured absorbance.

8. With light and concentration intertwined,
I describe the relation, clearly defined.

Absorbance and path length in the mix,

What's the law that gives you your fix?

Hint: I state that absorbance is directly proportional to concentration.

9. With bonds vibrating, I reveal the tune,

Molecular structures dance to my tune.

In the infrared, I find the way,

What technique am I? Can you say?

Hint: I identify molecular structures by analyzing vibrational transitions of bonds.

10. I measure the light that's beyond your sight,

In the UV and visible, I shine bright.

Analyzing samples, colors I find,

What's my name in this spectral bind?

Hint: I analyze how much UV and visible light a substance absorbs.

11. In magnetic fields, I reveal the truth,

Identifying structures, I'm the sleuth.

With nuclei spinning, I create a scene,

What technique am I? What do you glean?

Hint: I reveal molecular structure through the interaction of nuclear spins with magnetic fields.

12. I measure mass with precision true,

Fragments and ions reveal their due.

With each peak, a story unfolds,

What's this technique, if I may be so bold?

Hint: I separate ions based on their mass-to-charge ratio to identify compounds.

13. With inelastic scattering, I find the light,

Molecular vibrations come into sight.

In the visible, I provide a view,

What technique am I? Can you construe?

Hint: I use laser light to measure vibrational transitions in molecules.

14. In every color, I play a role,

Absorbing light, I'm on a stroll.

What part of the molecule am I, do you know,

That gives color to the light's flow?

Hint: I'm the part of a molecule responsible for its color due to light absorption.

15. In my presence, signals can sway,

Changing how compounds behave in a play.

What term describes how I alter the scene,

In spectra and signals, I'm keen?

Hint: I refer to how the choice of solvent can affect spectral measurements.

16. How well can you distinguish the peaks?
In the realm of spectra, clarity speaks.
Higher values give clearer sight,
What's this term that ensures things are right?
Hint: I measure the ability to separate closely spaced spectral features.

17. I measure amounts, not just the kind,
Finding concentrations, precise and blind.
In the lab, I bring the truth to light,
What's this analysis that makes things right?
Hint: I involve determining the quantity of an analyte in a sample.

18. I reveal what's there, without the number,
In colors and patterns, I make you wonder.
Identifying substances, I have my say,
What's this analysis that leads the way?
Hint: I identify the components of a mixture without quantifying them.

19. Before analysis, I make it neat,
Removing noise, making it complete.

Filtration, dilution, or extraction,

What's this step before the main action?

Hint: I involve the techniques used to prepare a sample for analysis.

20. With math and light, I find my place,

Analyzing bonds in a spectral space.

I gather all at once, not one by one,

What's my name in this analytical fun?

Hint: I use Fourier-transform techniques to obtain IR spectra rapidly.

Chapter 22

Chromatography

1. I'm the layer that stays, while others flow,

In my presence, the separation will grow.

What am I in this analytical race,

That helps to part components with grace?

Hint: I'm the fixed phase that interacts with the sample as it moves.

2. I'm the solvent that carries, moving with speed,

Transporting the sample to fulfill its need.

What's my name in this chromatographic dance,

As I help to separate in a fluid romance?

Hint: I'm the phase that moves through the stationary phase to carry the sample.

3. In a TLC plate, I show my worth,

Describing how far the spots traverse the earth.

With a simple ratio, I guide the way,

What's this factor that helps you portray?

Hint: I represent the distance traveled by a compound divided by the solvent front distance.

4. In a long tube, I separate with flair,

Packing materials help me do my share.

What technique am I, in layers so tall,

That divides mixtures, one and all?

Hint: I use a column packed with stationary phase to separate components in a liquid.

5. On a flat plate, I work with grace,

Revealing compounds in a small space.

With a solvent front, I'm quick and neat,

What am I in this analytical feat?

Hint: I'm a technique where a thin layer of adsorbent separates compounds based on their affinities.

6. In a vapor state, I analyze well,

Separation occurs, can you tell?

With a carrier gas, I find the right way,

What's this method that's here to stay?

Hint: I separate and analyze volatile compounds using a gas as the mobile phase.

7. Under pressure, I make my move,

Separating compounds with precision to prove.

In liquid form, I excel with speed,

What am I in this analytical creed?

Hint: I use high pressure to push the mobile phase through the stationary phase for fast separations.

8. In a state that's between gas and liquid I thrive,
Using supercritical fluids, I come alive.
What's this technique that's quite unique,
Separating compounds, the goal I seek?
Hint: I utilize supercritical fluids as the mobile phase for high efficiency in separations.

9. I define how well I can separate the peaks,
In graphs, I show if the method speaks.
What's this term that measures the clarity,
Of components in their spectral rarity?
Hint: I indicate the ability to distinguish between closely eluting compounds.

10. In two phases, I describe the fate,
How a solute divides, can you relate?
What's this measure that helps to explain,
The distribution of solutes in this domain?
Hint: I express the ratio of a compound's concentration in two immiscible solvents.

11. I change the composition, shift with grace,

Helping to separate in a dynamic space.

With varying conditions, I pave the way,

What's this technique that makes compounds sway?

Hint: I refer to the method of gradually changing the mobile phase composition during chromatography.

12. In the beginning, I play my part,

Introducing the sample, a vital start.

What's this step that sets the tone,

For the analysis that you'll own?

Hint: I refer to the process of injecting a sample into the chromatographic system.

13. I sense the compounds as they pass by,

Turning signals into data, oh my!

What's my role in this analytical scene,

To show you the presence of what might be seen?

Hint: I'm an instrument that measures and records the response of the separated components.

14. In a chromatogram, I set the stage,

A reference point that helps you gauge.

What's this line that runs so true,

Indicating where the signals accrue?

Hint: I'm the level on a chromatogram that represents the absence of analytes.

15. I'm the process of washing the sample away,
Separating compounds in a controlled display.
What's this term that signifies flow,
In the chromatographic show?

Hint: I refer to the process of carrying solutes through the stationary phase with the mobile phase.
16. I'm the goal of the chromatographic fight,
Dividing mixtures to bring clarity and light.
What's this term that describes the feat,
Of isolating components, oh so neat?
Hint: I describe the process of isolating different compounds within a mixture.

17. I'm the binding that takes place on the surface,
In chromatography, I serve with purpose.
What's this process that helps to attach,
Solutes to the phase, without a scratch?
Hint: I refer to the process where molecules adhere to the surface of the stationary phase.

18. In binding I trust, to separate and find,
Specific interactions of the molecular kind.

What's this method that uses a bait,

To capture the target, isn't it great?

Hint: I utilize specific interactions between a ligand and its target

molecule for separation.

19. In a world of size, I separate the small,

Larger ones pass through without a stall.

What's this technique that helps to define,

The molecules' sizes in this design?

Hint: I separate molecules based on their size, allowing smaller

molecules to elute last.

20. In the lab, I'm the careful plan,

Creating protocols for the analytical man.

What's this process that fine-tunes the way,

To achieve the best results day by day?

Hint: I refer to the systematic approach to create and optimize a

chromatographic method.

BIOCHEMISTRY

Chapter 23
Proteins

1. In chains, I come in many a hue,

Building blocks of life, I'm essential too.

Small but mighty, in proteins I play,

What am I, in your body's ballet?

Hint: I'm the smallest unit that makes proteins.

2. I link the units in a strong embrace,

Joining amino acids in a tight space.

A bond so special, I hold them tight,

What am I, in this protein flight?

Hint: I'm formed during protein synthesis.

3. In sequence I'm simple, a line so neat,

The order of my parts makes me complete.

Start with me first, then build on your way,

What structure am I, leading the play?

Hint: I'm the first level of protein structure.

4. I twist and turn in patterns so fine,

Alpha and beta, in shapes that align.

A helix or sheet, I'm quite the sight,

What am I, in this structural light?

Hint: I describe local folded structures.

5. I'm a 3D wonder, complex and grand,

Formed by interactions, not just by hand.

I give function to proteins, a shape to my role,

What am I, in the biochemistry scroll?

Hint: I refer to the overall folding of a protein.

6. Together we gather, multiple chains,

A complex formation where each one gains.

We function together, a team we create,

What structure am I, isn't it great?

Hint: I involve more than one polypeptide chain.

7. I speed up reactions, make processes flow,

Without me, life's pace would be slow.

Catalysts by nature, I'm key in the game,

What am I, in the biochemistry fame?

Hint: I'm crucial for metabolic reactions.

8. I'm the target that enzymes hold dear,

The molecules I bind bring the reaction near.

Together we dance in a biochemical play,

What am I, in this enzymatic ballet?

Hint: I'm the reactant that an enzyme acts upon.

9. In the heart of an enzyme, I'm where it's at,

The pocket that fits like a perfect hat.

Bind with precision, a match so divine,

What am I, in this molecular design?

Hint: I'm the region where substrate binding occurs.

10. I change my shape, I can twist and bend,

Regulating enzymes, I'm your friend.

Binding at a site that's not the main,

What am I, in the control chain?

Hint: I modify enzyme activity through binding elsewhere.

11. I'm the process that changes the structure so bold,

Heat or chemicals can make me unfold.

Proteins unravel, losing their might,

What am I, in the molecular fight?

Hint: I result in loss of protein function.

12. In the folding of proteins, I lend a hand,

Ensuring they fold just as they planned.

Helping them find their correct form,

What am I, in the protein norm?

Hint: I assist in proper protein folding.

13. After I'm made, I get a little twist,

Modifications that you shouldn't miss.

Phosphates and sugars, I wear like a crown,

What am I, in the cellular town?

Hint: I alter proteins after synthesis for function.

14. I'm the dance of polypeptides, a graceful spree,

Forming structures from chains, just wait and see.

A process that's vital, for function and form,

What am I, in this biological norm?

Hint: I determines a protein's shape and function.

15. I'm a soldier in the body, fighting off invaders,

Recognizing foes, I'm one of the traders.

Y-shaped and strong, I stand with pride,

What am I, in the immune tide?

Hint: I protect against pathogens and foreign substances.

16. I carry oxygen, red as a rose,

In your blood I travel, as everyone knows.

A globular protein, I'm vital indeed,

What am I, fulfilling this need?

Hint: I'm found in red blood cells.

17. I'm the glue that holds tissues with care,

In skin and in bones, I'm always there.

A fibrous protein, strong and secure,

What am I, in the structural allure?

Hint: I'm the most abundant protein in the body.

18. I'm the magic that speeds up a reaction so fast,

Lowering the energy needed, making it last.

A chemical dance, I'm crucial, you see,

What am I, in the world of chemistry?

Hint: I refer to the process of increasing reaction rates.

19. I sit on the surface, waiting to see,

Signals from outside come looking for me.

Binding with partners, I start the response,

What am I, in this biochemical dance?

Hint: I transmit signals into cells.

20. I'm the pathway through which messages flow,

From outside to in, I help cells to know.

A cascade of signals, I make things clear,

What am I, in the cellular sphere?

Hint: I'm the process of converting a signal into a response.

21. I'm an enzyme with a slightly different name,

Catalyzing the same reaction, but not the same game.

Variants of function, I show diversity,

What am I, in enzymatic universality?

Hint: I have different forms that perform the same function.

22. I'm a tag that marks proteins for fate,

Signaling for destruction, it's never too late.

A process of recycling, I help keep it clean,

What am I, in the cellular scene?

Hint: I'm involved in targeting proteins for degradation.

23. I'm a part of the protein, functional and neat,

Each domain has a role, making me complete.

In folds and in patterns, I do my job right,

What am I, in the protein's insight?

Hint: I'm a distinct structural unit of a protein.

24. I'm a regulator, binding to DNA,

Controlling the genes in a precise way.

Activating or silencing, I hold the key,

What am I, in gene therapy?

Hint: I influence gene expression.

25. I measure the masses, in a cloud I reside,

Identifying proteins, I'm the science guide.

Analyzing fragments with precision and grace,

What am I, in this analytical space?

Hint: I'm a technique used to identify molecules.

Chapter 24

Carbohydrates

1. Simple and sweet, I'm a single chain,

Glucose, fructose—my fame is plain.

The building blocks of carbs, I'm number one,

What am I, in this sugar run?

Hint: I'm the simplest form of carbohydrate.

2. Two of us join in a sweet embrace,

Sucrose and lactose, we're in the race.

A bond we form, then we're on our way,

What are we, in the sugary display?

Hint: I'm formed from two monosaccharides.

3. A few sugars linked, not too long,

Three to ten units, where we belong.

Found in fibers, we're not just for fun,

What are we, in this carbohydrate run?

Hint: I consist of a small number of sugar units.

4. Long chains of sugar, I'm complex and wide,

Starch and glycogen, I'm your energy guide.

Storage or structure, my roles are clear,

What am I, in this carbohydrate sphere?

Hint: I'm made of many monosaccharides.

5. I'm the bond that links, forming chains so neat,

Between sugar molecules, I'm a tasty treat.

A vital connection in carbohydrate lore,

What am I, that links sugars for sure?

Hint: I'm formed between two sugar molecules.

6. In plants, I'm stored, a source of delight,

A polysaccharide that's energy bright.

Found in potatoes and grains, I'm a must,

What am I, in the carbohydrate crust?

Hint: I'm a major energy storage form in plants.

7. In animals, I'm the storage king,

A branched polysaccharide, energy I bring.

In liver and muscle, I'm stored with care,

What am I, when energy's rare?

Hint: I'm the animal equivalent of starch.

8. In plant walls, I provide the strength,

A fiber so tough, I go to great lengths.

Indigestible for most, I'm still a friend,

What am I, on which plants depend?

Hint: I'm a structural polysaccharide in plants.

9. In arthropods and fungi, I make my stand,

A tough exoskeleton, I'm quite well planned.

Polysaccharide, but not from plants,

What am I, in the animal dance?

Hint: I'm similar to cellulose but found in animals and fungi.

10. I'm the process that turns food into fuel,

Breaking down sugars, I'm nobody's fool.

Energy from carbs, I bring to the scene,

What am I, in this metabolic machine?

Hint: I refer to how the body utilizes carbohydrates

11. In the cytoplasm, I break down the sweet,

Converting glucose to energy neat.

Ten steps of magic, I'm the first in line,

What am I, in this energy design?

Hint: I'm the metabolic pathway that converts glucose into pyruvate.

12. I'm the pathway that makes sugar anew,

From non-carbs, I'll build up for you.

In times of need, I'm ready to go,

What am I, in this metabolic show?

Hint: I'm the synthesis of glucose from non-carbohydrate sources.

13. I'm a metabolic route, branching away,

Producing NADPH for the cellular play.

Ribose-5-phosphate, I'm part of your tale,

What am I, in this metabolic trail?

Hint: I'm involved in the production of nucleotides and NADPH.

14. I'm a hormone that regulates your sugar's fate,

Lowering levels, I help you relate.

In the pancreas, I'm made with care,

What am I, in the glucose affair?

Hint: I help control blood sugar levels.

15. In response to low sugar, I'll raise the tide,

From the pancreas, I'm your glucose guide.

Promoting release of stored energy,

What am I, in the metabolic sea?

Hint: I act to increase blood sugar levels.

16. I'm a measure of glucose, critical to know,

Too high or too low can cause quite a show.

A balance so delicate, I keep you on track,

What am I, in the body's feedback?

Hint: I'm vital for energy homeostasis.

17. I'm the enzyme that starts the glucose spree,

Phosphorylating sugars to set them free.

With energy input, I pave the way,

What am I, in the glycolytic play?

Hint: I phosphorylate glucose to trap it in the cell.

18. A sugar so sweet, found in fruit and more,

I'm a monosaccharide that you can't ignore.

Processed in the liver, I play my part,

What am I, in the sugary heart?

Hint: I'm a common sugar found in many fruits.

19. In milk and dairy, I come to play,

A sugar so sweet in a different way.

Part of lactose, I'm linked with pride,

What am I, in the carbohydrate tide?

Hint: I'm a component of the sugar in milk.

20. A sugar alcohol, I'm sweet yet low,

Found in fruits, I help to balance the flow.

A role in metabolism, I'm often found,

What am I, in this sweet background?

Hint: I'm used as a sugar substitute.

21. A sugar of plants, I'm not too common,

In glycoproteins, I play the role of a summon.

Sweet and simple, I have my share,

What am I, in the sugar affair?

Hint: I'm a sugar involved in glycoprotein synthesis.

22. In seaweed and plants, I'm found in the mix,

A sugar that plays in cellular tricks.

Often attached to proteins, I help cells to bind,

What am I, in this carbohydrate grind?

Hint: I'm important in cell recognition.

23. I'm a combo of sugars and proteins or lipids,

In cell signaling, I help to keep lids.

Present on surfaces, I play a key role,

What am I, in this molecular goal?

Hint: I consist of carbohydrates linked to proteins or lipids.

24. I'm a protein that binds to sugar with grace,

Mediating interactions in this crowded space.

In plants and animals, I find my place,

What am I, in this carbohydrate chase?

Hint: I recognize and bind specific carbohydrates.

25. I'm the study of sugars, a field so vast,

Understanding their roles, present and past.

Carbohydrates matter, in health and disease,

What am I, in this scientific breeze?

Hint: I explore the structure and function of carbohydrates in biology.

Chapter 25

Lipids

1. I'm long and chain-like, with a carboxyl end,

Saturated or not, my properties blend.

Energy storage is where I excel,

What am I, can you tell?

Hint: I'm a building block of fats and oils.

2. Three fatty acids in a glycerol embrace,

In adipose tissue, I find my place.

A major form of stored energy, it's true,

What am I, can you construe?

Hint: I'm the main type of fat found in the body.

3 . With a hydrophilic head and tails that hide,

I form cell membranes, a protective guide.

A bilayer I make, so vital and grand,

What am I, in this cellular land?

Hint: I make up the structure of cell membranes.

4. Four fused rings, my structure's unique,

Hormones and cholesterol, I help you seek.

From testosterone to estrogen, I play my part,

What am I, a lipid with art?

Hint: I'm a type of lipid with a specific structure.

5. In membranes, I'm found, providing stability,
A precursor to hormones, I have versatility.
Too much of me can cause quite a fuss,
What am I, in this lipid discussion?

Hint: I'm important for membrane fluidity.

6. No double bonds in my long hydrocarbon chain,
Solid at room temperature, I'm often a gain.
Found in butter and meat, I'm rich and I'm bold,
What am I, in the lipid fold?
Hint: I'm typically solid at room temperature.

7. I have double bonds that make me bend,
Liquid at room temperature, on me you can depend.
Olive oil and avocados, I'm a healthy delight,
What am I, in the lipid fight?
Hint: I can be monounsaturated or polyunsaturated.

8. I'm modified to stay solid, though I'm not so great,
Linked to health issues, I tempt your fate.
Processed and hidden, I'm found in some snacks,
What am I, in this lipid act?
Hint: I'm formed through hydrogenation.

9. I carry lipids through the bloodstream with ease,

Chylomicrons and LDLs are part of my fees.

A mix of lipids and proteins, I'm essential for health,

What am I, in this lipid wealth?

Hint: I transport fats in the blood.

10. I'm a phospholipid that plays many roles,

In cell membranes, I help control.

A component of lecithin, I'm quite the find,

What am I, in this lipid bind?

Hint: I'm important for cellular signaling.

11. With three carbons, I form the backbone,

Of triglycerides, my fame has grown.

A simple alcohol, I'm sweet to taste,

What am I, in this lipid haste?

Hint: I'm the glycerin that links fatty acids.

12. A, D, E, and K, I'm essential you see,

In fatty tissues, I find my spree.

With lipids I travel, to keep you bright,

What am I, in the vitamin light?

Hint: I need fat for absorption in the body.

13. I'm a protective coating, smooth and fine,

Found on leaves and in bees, I'm truly divine.

Waterproofing surfaces, I'm quite the aid,

What am I, in this lipid parade?

Hint: I provide a protective barrier in nature.

14. I'm derived from fatty acids, a signaling crew,

Prostaglandins and leukotrienes, I'm part of the view.

Mediators of inflammation, I have my say,

What am I, in this lipid play?

Hint: I play roles in signaling and inflammation.

15. With my double bonds far apart, I'm quite the catch,

Found in fish and flaxseed, I'm a healthy match.

Heart-healthy and vital, I'm in high demand,

What am I, in this lipid strand?

Hint: I'm known for my anti-inflammatory properties.

16. Essential for growth, I'm found in seeds,

Sunflower and corn, I'm what your body needs.

Promoting inflammation, I have my say,

What am I, in this lipid fray?

Hint: I'm necessary for certain bodily functions.

17. I'm part of the membrane, a complex delight,

Sphingosine backbone, my structure's just right.

In the nervous system, I play a key role,

What am I, in this lipid goal?

Hint: I'm involved in cell recognition and signaling.

18. I'm formed in water, with lipids at my core,

A structure that helps fats to explore.

Transporting lipids, I'm quite the team,

What am I, in this lipid dream?

Hint: I facilitate the absorption of fats in the intestine.

19. With long chains and an alcohol end,

I'm used in cosmetics, I'm your friend.

A source of energy, I'm versatile too,

What am I, in this lipid view?

Hint: I can be found in various industrial applications.

20. I'm the structure that forms the cell's outer space,

With phospholipids layered, I provide a safe place.

Fluid and dynamic, I keep things in check,

What am I, in this cellular trek?

Hint: I'm the foundation of cellular membranes.

21. When glucose is scarce, I'm a fuel to find,

Produced from fatty acids, I'm one of a kind.

Used by the brain, I'm an energy source,

What am I, in this metabolic course?

Hint: I'm produced during fat metabolism.

22. I'm where fat is stored, a cushion for you,

Insulating the body, and energy too.

Brown and white, I have my share,

What am I, in this lipid affair?

Hint: I serve as energy storage in the body.

23. From lipids I'm made, signaling away,

Regulating functions, day by day.

Cortisol and testosterone, I'm quite the team,

What am I, in this hormonal dream?

Hint: I play crucial roles in body regulation.

24. In processed foods, I'm what you might see,

Replacing fats to cut calories.

Mimicking flavor, I help in the mix,

What am I, in this dietary fix?

Hint: I'm used to reduce fat content in food.

25. I'm the study of lipids, their roles and more,

Understanding how they function is what I explore.

In health and disease, I have much to say,

What am I, in this scientific play?

Hint: I examine lipid profiles in biological systems.

Chapter 26
Nucleic Acids

1. I'm the building block, so small and neat,

A sugar, a phosphate, and a base complete.

DNA and RNA, I help to compose,

What am I, in this genetic prose?

Hint: I'm the monomer of nucleic acids.

2. Double helix twisted, I carry your code,

The blueprint of life, on me it's bestowed.

Adenine, thymine, cytosine, guanine too,

What am I, holding secrets of you?

Hint: I store genetic information in cells.

3. Single-stranded and versatile, I'm often in play,

mRNA, tRNA, I help in the fray.

From DNA I'm transcribed, a messenger bold,

What am I, with stories to be told?

Hint: I'm essential for protein synthesis.

4. I'm one of the bases, a purine, you see,

Pairing with thymine, I'm part of the spree.

Energy carrier, I'm ATP's best friend,

What am I, in this genetic blend?

Hint: I'm one of the nucleobases in DNA and RNA.

5. A pyrimidine base, I pair with A,

In DNA I dwell, helping code the way.

Essential for structure, I'm one of the four,

What am I, at the genetic core?

Hint: I'm found only in DNA, not RNA.

6. I'm another base, pairing with G,

In DNA and RNA, I'm part of the spree.

A pyrimidine like thymine, I play my part,

What am I, in this molecular art?

Hint: I'm a component of both DNA and RNA.

7. I'm a purine base, with cytosine I pair,

In the nucleic acids, I'm found everywhere.

Essential for coding, I'm part of the game,

What am I, in this genetic frame?

Hint: I'm present in both DNA and RNA.

8. I'm a base that's found in RNA, it's true,

Replacing thymine, I'm a part of the crew.

Pairing with adenine, I take my stand,

What am I, in this nucleic strand?

Hint: I'm unique to RNA and not found in DNA.

9. I'm the process that makes DNA anew,

Copying the strands, it's what I do.

Semi-conservative, I'm key to the game,

What am I, in this genetic frame?

Hint: I'm how DNA is duplicated.

10. I'm the first step in making a gene,

From DNA to RNA, my work is seen.

RNA polymerase, I guide in this task,

What am I, in this nucleic flask?

Hint: I'm the process of synthesizing RNA from DNA.

11. I'm the process that turns RNA to protein,

Ribosomes and tRNA, I'm part of the scene.

Codons and anticodons, I help to align,

What am I, in this biochemical design?

Hint: I'm how proteins are synthesized from mRNA.

12. I carry the code from the nucleus wide,

A template for proteins, I'm a cellular guide.

Transcribed from DNA, I'm the message in hand,

What am I, in this genetic band?

Hint: I convey genetic information from DNA to ribosomes.

13. I bring the amino acids to the ribosome site,

Anticodons matching codons, I make it right.

A key player in translation, I'm essential to all,

What am I, in this protein call?

Hint: I transfer specific amino acids during protein synthesis.

14. I'm part of the ribosome, making proteins go,

Catalyzing reactions, I'm the worker, you know.

A structural component, I'm essential and grand,

What am I, in this cellular band?

Hint: I'm a key component of ribosomes.

15. Three bases in sequence, I code for a fate,

An amino acid's identity, I help to create.

In mRNA, I'm the language of life,

What am I, in this genetic strife?

Hint: I'm a sequence of three nucleotides that codes for an amino acid.

16. I'm the counterpart to a codon, it's true,

On tRNA I'm found, pairing with you.

Ensuring the right amino acids are there,

What am I, in this genetic affair?

Hint: I'm a sequence of three nucleotides that pairs with a codon in mRNA.

17. I'm made of DNA, tightly coiled and wrapped,

In the nucleus, I'm stored, neatly mapped.

Carrying genes, I'm essential to life,

What am I, in this genetic strife?

Hint: I'm a structure that carries genetic information.

18. I'm a protein that helps DNA to bind,

Forming nucleosomes, a structure refined.

Packing it tightly, I play my role,

What am I, in this genetic scroll?

Hint: I'm a protein that helps in DNA packaging.

19. I'm a segment of DNA, coding a trait,

From eye color to height, I determine fate.

The basic unit of heredity, I'm found in the strand,

What am I, in this biological land?

Hint: I'm a specific sequence of DNA that codes for a protein.

20. I'm the variants of a gene, don't you see?

Dominant or recessive, I play with glee.

Inheriting traits, I add to the mix,

What am I, in this genetic fix?

Hint: I represent different forms of the same gene.

21. I'm a change in the DNA sequence, it's true,

Sometimes harmful, sometimes new.

Driving evolution or causing disease,

What am I, in this genetic tease?

Hint: I'm a change that occurs in the nucleotide sequence.

22. I'm a non-coding region, a segment of DNA,

Spliced out of mRNA, I quietly stay.

Though I don't code for proteins, I have my say,

What am I, in this genetic ballet?

Hint: I'm a sequence of DNA that is not translated into protein.

23. I'm the coding sequence that stays in the game,

Translated to protein, I take my aim.

Spliced together, I play my part,

What am I, in this genetic art?

Hint: I'm a segment of DNA that codes for a protein.

24. I'm the process of turning genes into action,

Regulating when and how, I'm a big attraction.

From transcription to translation, I hold the key,

What am I, in this cellular spree?

Hint: I'm how information from a gene is used to synthesize a functional gene product.

25. I'm a tool for editing genes, cutting with ease,

Revolutionizing science, aiming to please.

With Cas9 I work, making changes profound,

What am I, in this genetic ground?

Hint: I'm a technology used for precise gene editing.

ADVANCED TOPICS

Chapter 27
Quantum Chemistry

1. In waves and particles, I reside,

My dance unseen by human eyes.

Probability clouds are where I hide,

What am I, this quantum prize?

Hint: I'm the building block of atoms, negatively charged.

2. Discrete and jumping, never in-between,

My levels are strict, not a continuous scene.

Absorption and emission, my spectral sheen,

Which quantum concept am I, so keen?

Hint: Planck's constant is key to understanding me.

3. Spinning up or spinning down,

Half-integer values all around.

Paired or unpaired, I abound,

In atoms and molecules, I'm found.

Hint: I'm an intrinsic angular momentum.

4. Four numbers describe my state,

In atomic orbitals, I create.

No two alike, it's quantum fate,

Which principle do I relate?

Hint: I prevent electrons from occupying the same quantum state.

5. Wave functions squared, that's my game,

Chances of finding electrons, I name.

Born's my father, in physics fame,

What interpretation bears my name?

Hint: I give physical meaning to the wave function.

6. Uncertainty reigns in my domain,

Position and momentum, a blurry refrain.

The more you know one, the less the other's gain,

Whose principle causes this quantum strain?

Hint: A German physicist's name is associated with me.

7. S, p, d, and f, I'm known by these,

Shapes in space, electrons please.

From simple spheres to complex geometries,

What quantum concept do I seize?

Hint: I describe the probability distribution of electrons around the nucleus.

8. Two slits I pass, yet patterns emerge,

Particle or wave, my nature does surge.

Observer effects cause realities to merge,

Which experiment am I, on physics' verge?

Hint: I demonstrate wave-particle duality.

9. Schrödinger's equation, I satisfy,

Describing particles as time goes by.

Complex-valued function, don't ask why,

What mathematical tool do I supply?

Hint: I'm the fundamental equation of quantum mechanics.

10. Quantum tunneling is my trick,

Through barriers classical physics can't pick.

Alpha decay, my cosmic kick,

What phenomenon makes particles tick?

Hint: I allow particles to pass through energy barriers.

11. Bosons or fermions, I classify,

Half-integer or integer spin, I specify.

Particle behavior, I clarify,

Which quantum rule do I signify?

Hint: I connect a particle's spin to its statistical behavior.

12. Nodes and antinodes, I display,

Electron probability, I convey.

In atoms and molecules, I hold sway,

What visual tool am I, you'd say?

Hint: I'm a graphical representation of electron distribution.

13. Superposition is my state of grace,

Multiple states in a single space.

Collapse upon measurement, leave no trace,

What quantum concept do I embrace?

Hint: Schrödinger's cat is a famous thought experiment about me.

14. Entangled pairs, I correlate,

Spooky action at a distance, I create.

Einstein's worry, a quantum trait,

What phenomenon do I relate?

Hint: I connect particles regardless of their separation.

15. Aufbau, Hund, and Pauli combined,

Electron configurations, I help find.

Building up orbitals, all aligned,

What principle keeps atoms well-designed?

Hint: My name means "building up" in German.

16. Resonance structures, I explain,

Electron delocalization, my domain.

Benzene's stability, I maintain,

What concept clarifies this electron train?

Hint: I show how electrons are shared across a molecule.

17. HOMO and LUMO, I define,

Reactivity and spectra, I refine.

Electron transfers, I align,

What orbital theory do I outline?

Hint: I focus on the highest occupied and lowest unoccupied molecular orbitals.

18. Singlet or triplet, I dictate,

Photochemistry, I orchestrate.

Electron spins, I correlate,

What quantum state do I create?

Hint: I determine how electrons are paired in an atom or molecule.

19. Vibrational levels, I quantize,

Molecular motion, I comprise.

Infrared spectra, I revise,

What quantum model do I devise?

Hint: I'm often represented by a parabolic potential energy curve.

20. Exchange and correlation, I address,

Electron-electron interactions, I finesse.

Beyond Hartree-Fock, I progress,

What advanced method do I express?

Hint: I'm a computational method that models electron density.

Chapter 28
Surface Chemistry

1. I'm a force that's small, yet I bind them all,

On surfaces where molecules crawl.

What am I?

Hint: Think of tiny attractions.

2. I spread with ease, on solids I please,

Reducing tension, I'm sure to appease.

What am I?

Hint: I make things wetter.

3. I'm a layer so thin, just one atom within,

On metals I sit, with electrons I spin.

What am I?

Hint: I'm a chemist's monolayer dream.

4. I'm a process so neat, where gases and solids meet,

On surfaces I stick, in cleanups I'm quick.

What am I?

Hint: I cling to surfaces from the air.

5. I'm a point of change, where phases rearrange,

Three states converge, at temperatures strange.

What am I?

Hint: I'm a triple threat in phase diagrams.

6. I'm an angle acute, that droplets salute,

On surfaces I lie, wettability I denote.

What am I?

Hint: I measure how drops meet surfaces

7. I'm a layer so fine, of molecules in line,

On water I float, one molecule high.

What am I?

Hint: I'm named after an American scientist.

8. I'm a force that's strong, to surfaces I belong,

Pulling liquids up tubes, against gravity's song.

What am I?

Hint: I make water climb.

9. I'm a state of matter, neither solid nor liquid chatter,

On surfaces I spread, molecules barely tethered.

What am I?

Hint: I'm two-dimensional and fluid.

10. I'm a phenomenon grand, where liquids expand,

Across surfaces they land, in a layer so bland.

What am I?

Hint: I make drops become films.

11. I'm a measure of might, of a surface's bite,

Energy to create, new areas I state.

What am I?

Hint: I quantify surface creation energy.

12. I'm a graph so fine, of pressure versus confine,

For gases on solids, my curves intertwine.

What am I?

Hint: I show how gases stick as pressure rises.

13. I'm a layer with charge, where ions barge,

Near electrodes I form, potentials transform.

What am I?

Hint: I'm named after a German physicist.

14. I'm a process reversed, from surfaces dispersed,

Molecules break free, from their solid spree.

What am I?

Hint: I'm adsorption's opposite.

15. I'm a state of repel, where drops round and swell,

On surfaces they bead, not spreading indeed.

What am I?

Hint: Water on lotus leaves demonstrates me.

16. I'm an effect so small, making rough surfaces crawl,

With temperature's sway, particles find their way.

What am I?

Hint: I'm named after a Russian scientist.

17. I'm a model of bind, for gases you'll find,

One layer I predict, on surfaces strict.

What am I?

Hint: I assume one molecule thickness.

18. I'm a force electron, from surfaces I'm drawn,

In vacuum I race, from metals with grace.

What am I?

Hint: I emit electrons from surfaces.

19. I'm a process of change, crystal shapes I arrange,

On surfaces I grow, in patterns that show.

What am I?

Hint: I build crystals layer by layer.

20. I'm a coefficient of drag, where fluids and solids snag,

No slip do I allow, at interfaces I vow.

What am I?
Hint: I describe fluid-solid stickiness.

Chapter 29
Polymer Chemistry

1. I'm a chain so long, of units strong,

Repeated patterns, where molecules belong.

What am I?

Hint: I'm the essence of plastic.

2. I split and I join, in a dance so fine,

Equilibrium's my goal, in polymer's design.

What am I?

Hint: I balance making and breaking.

3. I'm a process of growth, where monomers oath,

Step by step, the chain's unclothe.

What am I?

Hint: I build polymers one unit at a time.

4.   I branch and I spread, like a tree overhead,

My polymer form, is far from uniform.

What am I?

Hint: I'm not a straight chain.

5. I'm a number so grand, of units at hand,

In a polymer chain, my value's not plain.

What am I?

Hint: I quantify polymer size.

6. Two types combined, in one chain entwined,
A copolymer's grace, with units interlaced.
What am I?
Hint: I mix monomers in one polymer.

7. I'm soft then I'm hard, as temperature's guard,
A polymer's transition, with molecular ambition.
What am I?
Hint: I mark a key thermal change.

8. I catalyze with zeal, make polymers real, amed for a Nobel's
seal, Ziegler-Natta's deal.
What am I?
Hint: I help make polyolefins.

9. I'm a polymer solution, in thermal evolution,
As heat rises high, from solution I fly.
What am I?
Hint: I separate with heat.

10. I'm a radical's friend, where polymers extend,
Chain reaction's my game, growing polymers my aim.

What am I?

Hint: I use free radicals to grow.

11. I'm a force so strong, where polymer chains belong,
Crossing paths with grace, in a network's embrace.
What am I?
Hint: I link polymer chains.

12. I'm a polymer's core, with tacticity in store,
Arranged just so, stereoregular I grow.
What am I?
Hint: I describe spatial arrangement.

13. Initiating with might, polymer growth I ignite.
What am I?
Hint: I kick-start the reaction.

14. I'm a polymer blend, where properties transcend,
Mixed but not bonded, new materials responded.
What am I?
Hint: I combine polymers physically.

15. I'm a chain transfer's art, where growth may restart,
To agent I leap, new polymer to reap.
What am I?

Hint: I switch growing chains.

16. I'm a polymer's end, where function can append,

Reactive and free, customizing with glee.

What am I?

Hint: I'm the polymer's reactive tip.

17. I'm a polymer's test, of time I protest,

Breaking bonds with ease, as oxygen seizes.

What am I?

Hint: I age polymers with air.

18. I'm a method so living, control I am giving,

Polymers grow neat, till told to complete.

What am I?

Hint: I grow polymers with precision.

19. I'm a polymer's measure, of molecular treasure,

Averaged with care, my distribution's fair.

What am I?

Hint: I represent typical polymer size.

20. I'm a change so profound, where polymers are bound,

With heat or light's sway, I cure polymers' way.

What am I?

Hint: I harden liquid polymers.

Chapter 30
Green Chemistry

1   I'm a principle so green, in chemistry I'm keen,

Prevention's my game, waste I aim to tame.

What am I?

Hint: I'm first among twelve.

2. I maximize matter, in reactions I flatter,

Efficiency's my friend, to products I tend.

What am I?

Hint: I use everything I can.

3. I'm a solvent so neat, water can't compete,

Super critical and clean, in extractions I'm keen.

What am I?

Hint: I'm CO2 in a special state.

4. I catalyze with ease, Mother Nature I please,

Enzymes are my tool, in green chem I rule.

What am I?

Hint: I use nature's catalysts.

5. I'm energy's friend, on sun I depend,

Reactions I drive, helping green chem thrive.

What am I?

Hint: I harness solar power.

6. I'm a metric so true, impact I review,

Efficiency I measure, for chemistry's treasure.

What am I?

Hint: I quantify greenness.

7. I'm a reaction so swift, at room temp I'm adrift,

No heat do I need, green chem I heed.

What am I?

Hint: I work without warming.

8. I'm a source renewable, to petrol unbeatable,

From plants I derive, green fuels I contrive.

What am I?

Hint: I'm a biological alternative.

9. I'm a process design, where waste I malign,

From cradle to grave, resources I save.

What am I?

Hint: I consider the full lifecycle.

10. I'm a catalyst's pride, on surface I ride,

Reactions I speed, less metal I need.

What am I?

Hint: I'm a thin, active layer.

11. I'm a solvent ionic, to green chem symphonic,

At room temp I'm liquid, my uses are vivid.

What am I?

Hint: I'm a salt that flows.

12. I'm a reaction's delight, microwaves my might,

Faster and cleaner, energy I'm leaner.

What am I?

Hint: I cook reactions quickly.

13.   I'm a chemical's fate, in nature I translate,

Quickly I degrade, no pollution is made.

What am I?

Hint: I break down easily.

14. I'm a feed so stock, to petroleum a shock,

Renewable and green, in plastics I'm seen.

What am I?

Hint: I'm plant-based plastic source.

15. I'm a principle bold, toxicity I withhold,

Safer chemicals I preach, benign design I teach.

What am I?

Hint: I'm the 4th green principle.

16. I'm a reaction's trend, where solvents I end,

Solvent-free I go, green chemistry I show.

What am I?

Hint: I mix without dissolving.

17. I'm a source of good, from nature's neighborhood,

Medicines I inspire, green chem's desire.

What am I?

Hint: I find drugs in forests.

18. I'm a process real-time, efficiency's paradigm,

Monitoring reactions, preventing infractions.

What am I?

Hint: I watch as it happens.

19. I'm a metric of green, twelve principles I glean,

A star for each one, green progress I've won.

What am I?

Hint: I rate on a dozen points.

20. I'm a change catalyzed, by light energized,

Clean and precise, nature's own device.

What am I?

Hint: I use light and life.

BONUS CHAPTER

Amazing Facts From Chemistry

1. The world's smallest periodic table was etched on a human hair.

Yes, Professor Martyn Poliakoff has had the periodic table engraved on one of the hairs from his head.

2. Glass is actually a liquid that flows very slowly. Ancient windows are thicker at the bottom due to this flow.

3. There's enough gold dissolved in the world's oceans to give every person on Earth 9 pounds of gold.

4. Helium is the only element discovered outside of Earth before it was found on Earth. It was first detected in the sun's spectrum.

5. Diamonds aren't rare; they're actually the most common gemstone. Their price is kept artificially high by controlling supply.

6. Water can boil and freeze at the same time. This occurs at 01°C and 73 pascals of pressure, known as the triple point of water.

7. The longest chemical name has 189,819 letters and takes about three and a half hours to pronounce. It is the chemical name for the largest known protein titin.

8. If you mix equal amounts of water and alcohol, the volume of the resulting solution will be less than the sum of the two initial volumes. This is due to the intermolecular attraction between water and ethanol.

9. A rubber tire is actually one single giant molecule.

10. Chocolate contains a chemical called theobromine, which is poisonous to dogs and cats.

11. Gallium is a metal that can melt in your hand because its melting point is only about 6°F (8°C).

12. Aerogel, nicknamed "frozen smoke," is the world's lightest solid material, composed of 8% air.

13. Bismuth is the most naturally diamagnetic element, meaning it's repelled by both poles of a magnet.

14. Francium is the rarest naturally occurring element on Earth's crust, with only a few grams existing at any given time. Estimates suggest that only 30 g of it is present in the whole of Earth's crust.

15. Sulfur hexafluoride (SF6) is a gas so dense that you can float lightweight objects on it like a liquid.

16. Teflon, used in non-stick cookware, was discovered by accident when a researcher was trying to make a new refrigerant.

17. Hydrofluoric acid is so corrosive that it can dissolve glass and most metals.

18. The longest-living radioactive isotope is tellurium-128, with a half-life of 160 trillion years.

19. Buckminsterfullerene (C60), also known as a "buckyball," is a spherical molecule that looks like a soccer ball.

20. The Mpemba effect is the observation that, under certain conditions, warmer water can freeze faster than colder water.

21. Serotonin, often called the "happy chemical," is also found in bananas.

22. The Periodic Table's creator, Dmitri Mendeleev, used to make suitcases in his spare time.

23. Thallium is so poisonous that it was once used as rat poison but has since been banned for consumer use.

24. J and Q are the only letters not found in the periodic table because they do not occur in either element symbols or element names.

25. The element californium is so radioactive that it glows in the dark.

26. Copper doorknobs are self-disinfecting. The ions in the metal have antimicrobial properties.

27. Ice can burn. Yes, when in extreme cold, water can freeze so quickly that it traps methane gas inside, forming what is known as methane hydrate. This strange form of ice can actually catch fire when brought into a warmer environment.

28. The artificial sweetener sucralose (Splenda) was discovered when a scientist misheard an instruction to "test" a compound as an instruction to "taste" it.

29. Europium, a rare earth element, is used to make euro banknotes glow under UV light as an anti-counterfeiting measure.

30. Bose-Einstein condensates, a state of matter formed at extremely low temperatures, can slow light to less than 60 kilometers per hour.

31. The chemical element copernicium is so unstable that only a few atoms of it have ever been created, and they lasted just a few seconds before decaying.

32. The compound with the worst smell ever recorded is thioacetone. It can cause nausea, vomiting, and unconsciousness from miles away.

33. Chirality in organic molecules can drastically change their properties. For instance, one enantiomer of limonene smells like oranges, while the other smells like lemons.

34. Organic chemists have synthesized a molecule called olympicene, whose structure resembles the Olympic rings.

35. The compound resveratrol, found in red wine, has been shown to have anti-aging properties in certain organisms.

36. Indigo, the compound that gives blue jeans their color, was originally extracted from plants but is now synthesized on an industrial scale.

37. Benzene's structure stumped chemists for years until Friedrich August Kekulé reportedly dreamed of a snake biting its own tail, inspiring the cyclic structure we know today.

38. The artificial sweetener saccharin was discovered when a chemist forgot to wash his hands before eating dinner and noticed his food tasted sweet.

39. Indium tin oxide is both electrically conductive and optically transparent, making it crucial for touchscreen devices.

40. The compound boron nitride can be as hard as diamond in its cubic form, and as soft as graphite in its hexagonal form.

41. Tungsten, in pure form, has the highest melting point, i.e., 3422 °C or 6192 °F.

42. Though we think of oxygen as a colorless gas, when it's cooled to a liquid state, it turns into a pale blue liquid. Back To Riddles

43. Bananas contain potassium-40, a naturally occurring isotope that is radioactive. Though it's not harmful, it's interesting that foods like bananas can emit radiation.

44. While sulfuric acid is strong, carborane acid is 100,000 times stronger, yet it's so gentle that it doesn't corrode skin.

45. Since peanut butter is rich in carbon, scientists have been able to create diamonds from it under extreme heat and pressure conditions in a lab.

46. Ethanol is both a fuel and a beverage. Ethanol (C_2H_5OH), the type of alcohol found in alcoholic drinks, is also used as a biofuel. However, while it's safe to drink in moderation,

consuming fuel-grade ethanol is toxic because it contains additives to prevent ingestion.

47. Spider silk is a natural polymer. It is made of proteins, which are organic polymers. It's stronger than steel by weight and has a wide range of potential applications in medicine and materials science.

48. Chlorine gas is highly toxic, but chlorine ions (chloride) are essential for life and are found in table salt (sodium chloride).

49. Avogadro's number (6.023×10^{23}) is the number of particles in one mole of a substance. This number is so large that if you had Avogadro's number of unpopped popcorn kernels, they would cover the Earth to a depth of about 9 miles.

50. Sonoluminescence is a fascinating phenomenon where small gas bubbles in a liquid emit a brief flash of light when exposed to intense sound waves, momentarily reaching temperatures hotter than the surface of the sun.

ANSWERS

BASICS

Chapter 1

Atoms and Elements

1. Atom
2. Hydrogen
3. Electron
4. Neutron
5. Proton
6. Atomic number
7. Electron shells (or energy levels)
8. Octet rule
9. Mass number
10. Isotopes
11. Valence electrons
12. Lone pair electrons
13. Nuclear attraction
14. Atomic radius
15. Atomic orbital
16. Stoichiometric coefficient
17. Density
18. Electron orbit
19. Energy level
20. Radioactive decay
21. Plum pudding model
22. Electrostatic repulsion
23. Electron configuration
24. Electron affinity
25. Ion
26. Strong nuclear force
27. Atomic mass unit (amu)
28. Heisenberg uncertainty principle

29. Ionization energy
30. Aufbau principle
31. Electromagnetic force
32. Unpaired electron
33. Nuclear spin
34. Atomic emission
35. Quantum numbers
36. Effective nuclear charge
37. Electron shell
38. s orbital
39. p orbitals
40. Thomson's atomic model
41. Atomic empty space
42. Group (or family)
43. Electropositivity
44. Spin quantum number
45. Chemical symbol
46. Elementary charge
47. Electron shielding
48. Pauli exclusion principle
49. Metallic bond
50. d orbitals
51. Electronic transition
52. Covalent radius
53. Spin statistics
54. Octet configuration
55. Azimuthal quantum number (Angular momentum quantum number)
56. Ground state
57. Orbital diagram
58. Magnetic quantum number
59. Polarity
60. Bohr model
61. Atomic weight
62. Zero-point energy
63. Nuclear magnetic resonance

23. Helium
24. Phosphorus
25. Mercury
26. Sodium
27. Nickel
28. Argon
29. Samarium
30. Tantalum
31. Platinum
32. Iodine
33. Potassium
34. Copper
35. Vanadium
36. Fluorine
37. Krypton
38. Gadolinium
39. Arsenic
40. Cesium
41. Manganese
42. Aluminum (Aluminium)
43. Zinc
44. Bromine
45. Erbium
46. Lead
47. Radium
48. Tungsten
49. Neon
50. Nitrogen
51. Palladium
52. Chlorine
53. Oxygen
54. Gold
55. Antimony
56. Rubidium
57. Rhenium
58. Xenon

Chapter 3

Chemical Bonds

1. Covalent Bond
2. Electrons
3. Electron Pair
4. Bond Pair
5. Lone Pair
6. Shared Electrons
7. Nonpolar Covalent Bond
8. Polar Covalent Bond
9. Bond Length
10. Bond Energy
11. Bond Angle
12. Sigma Bond (σ)
13. Pi Bond (π)
14. Single Bond
15. Double Bond
16. Triple Bond
17. Valence Electrons
18. Lewis Structure
19. Octet Rule
20. Electronegativity
21. Dipole Moment
22. Molecular Orbital
23. Overlap of Orbitals
24. Hybridization
25. sp Hybridization
26. sp^2 Hybridization
27. sp^3 Hybridization

28. VSEPR Theory
29. Molecular Geometry
30. Covalent Radius
31. Bond Dissociation Energy
32. Covalent Compound
33. Molecule
34. Resonance Structures
35. Delocalized Electrons
36. Formal Charge
37. Coordinate Covalent Bond
38. Partial Charges
39. Bond Polarity
40. Electron Density
41. Polar Molecule
42. Nonpolar Molecule
43. Bond Strength
44. Dative Bond
45. Molecular Symmetry
46. Homonuclear Molecule
47. Heteronuclear Molecule
48. Intermolecular Forces
49. London Dispersion Forces
50. Hydrogen Bonding

Chapter 4

States of Matter

1. States of matter
2. Solid
3. Liquid
4. Gas
5. Plasma
6. Melting
7. Evaporation
8. Triple point

9. Sublimation
10. Latent heat of fusion
11. Supercritical fluid
12. Deposition
13. Boiling point
14. Melting point
15. Latent heat of condensation (or freezing)
16. Diffusion
17. Intermolecular forces
18. Viscosity
19. Surface tension
20. Gas pressure
21. Colloid (or gel)
22. Deformation
23. Elasticity
24. Plasticity
25. Brownian motion
26. Continuous phase transition
27. Hardness
28. Coagulation
29. Ductility
30. Surface film
31. Capillary action
32. Amorphous solid
33. Desublimation (or deposition)
34. Superfluid
35. Degenerate matter
36. Condensation
37. Sublimation point
38. Liquid crystal
39. Freezing (or solidification)
40. Bose-Einstein condensate
41. Effervescence (or degassing)
42. Colloid
43. Plastic deformation
44. Degenerate matter

45. Cooling
46. Phase transition (solid-solid)
47. Compression
48. Vapor-liquid mixture
49. Coalescence
50. Electron sea (or electron gas)
51. Comminution (or grinding)
52. Melting point (or freezing point)

INORGANIC CHEMISTRY

Chapter 5

Acids and Bases

1. Arrhenius Theory
2. Bronsted-Lowry Theory
3. Lewis Acid-Base Theory
4. pH and pOH Scale
5. Conjugate Acid-Base Pairs
6. Strong Acids
7. Weak Acids
8. Strong Bases
9. Weak Bases
10. Acid-Base Titration
11. Buffer Solutions
12. Acidic Oxides
13. Basic Oxides
14. Amphoteric Oxides
15. Neutral Oxides
16. Common Ion Effect
17. Autoionization of Water
18. Kw (Ionization Constant of Water)
19. Hydronium Ion (H_3O^+)
20. Hydroxide Ion (OH^-)
21. Acid Dissociation Constant (Ka)
22. Base Dissociation Constant (Kb)

Chapter 6

Redox Reactions

4. Reducing Agent
5. Redox Reaction
6. Oxidation Number
7. Disproportionation
8. Comproportionation
9. Half-Reaction Method
10. Electrochemical Cell
11. Galvanic Cell
12. Electrolytic Cell
13. Anode
14. Cathode
15. Salt Bridge
16. Electron Transfer
17. Oxidation State
18. Standard Electrode Potential
19. Nernst Equation
20. Electrochemical Series
21. Voltaic Cell
22. Electrolysis
23. Faraday's Laws of Electrolysis
24. Electrode Potential
25. Cell Potential (Ecell)
26. Overpotential
27. Corrosion
28. Rusting of Iron
29. Electroplating
30. Fuel Cell
31. Hydrogen Economy
32. Redox Titration
33. Potentiometric Titration
34. Winkler Method (Dissolved Oxygen Analysis)
35. Iodometric Titration
36. Permanganometry
37. Dichromate Titration
38. BOD (Biochemical Oxygen Demand)
39. COD (Chemical Oxygen Demand)

40. Electrochemical Equilibrium
41. Primary Battery
42. Secondary Battery
43. Lead-Acid Battery
44. Nickel-Cadmium Battery
45. Lithium-Ion Battery
46. Zinc-Carbon Battery
47. Alkaline Battery
48. Electron Flow
49. Cell Notation
50. Half-Cell
51. Standard Hydrogen Electrode (SHE)
52. Reference Electrode
53. Redox Potential
54. Latimer Diagram
55. Frost Diagram
56. Pourbaix Diagram
57. Charge Transfer Complex
58. Redox Couple
59. Oxidation of Metals
60. Reduction of Metals
61. Electrode Potential
62. Catalytic Converter
63. Hydrogen Peroxide (H_2O_2)
64. Chlorine Gas (Cl_2)
65. Fluorine (F_2)
66. Disproportionation Reaction
67. Standard Electrode Potential ($E°$)
68. Redox Cycle
69. Galvanic Corrosion
70. Redox Reaction in Photosynthesis
71. Redox Reaction in Respiration
72. Fehling's Test
73. Benedict's Test
74. Rusting of Iron
75. Passivation

Chapter 7

Coordination Compounds

Chapter 8

Nuclear Chemistry

18. Nuclear Fusion Reactor
19. Thermal Neutron
20. Mass Defect
21. Binding Energy
22. Photon
23. Nuclear Medicine
24. Tracer
25. Dosimetry
26. Nuclear Waste
27. Radiopharmaceuticals
28. Decay Series
29. Radiological Hazard
30. Geiger Counter
31. Becquerel
32. Curie
33. Sievert
34. Radionuclide
35. Neutron Capture
36. Thermonuclear Reaction
37. Reactor Coolant
38. Fission Products
39. Radiation Shielding
40. Nuclear Physics
41. Criticality
42. Neutrino
43. Spontaneous Fission
44. Fuel Rod
45. Control Rod
46. Ionization
47. Electron Capture
48. Beta Decay
49. Radiotoxicity
50. Nuclear Chain Reaction

ORGANIC CHEMISTRY

Chapter 9

Concepts in Organic Chemistry

1. Electronegativity
2. Polar Bond
3. Nonpolar Bond
4. Dipole Moment
5. Resonance
6. Delocalization
7. Inductive Effect
8. Hyperconjugation
9. Aromatic Stabilization
10. Ring Strain
11. Van der Waals Forces
12. Hydrogen Bonding
13. Solvation
14. Coordination
15. Lewis Structure
16. Molecular Orbital Theory
17. Valence Bond Theory
18. Hybridization
19. Sigma Bond
20. Pi Bond
21. Aromatic Compounds
22. Alkyl Halides
23. Alkyl Sulfides
24. Alkyl Amines
25. Organometallic Compounds
26. Synthesis Planning
27. Functional Group Interconversion
28. Mechanistic Rationale
29. Reaction Kinetics
30. Chemical Equilibrium

Chapter 10

Hydrocarbons

1. Alkane
2. Alkene
3. Alkyne
4. Aromatic
5. Saturated
6. Unsaturated
7. Isomer
8. Structural Isomer
9. Geometric Isomer
10. Cis-Isomer
11. Trans-Isomer
12. Hydrocarbon Chain
13. Linear Hydrocarbon
14. Branched Hydrocarbon
15. Cycloalkane
16. Polycyclic Hydrocarbon
17. Aliphatic
18. Aromaticity
19. Benzene
20. Toluene
21. Xylene
22. Naphthalene
23. Polycyclic Aromatic Hydrocarbon (PAH)
24. Combustion
25. Hydrocarbon Oxidation
26. Cracking
27. Hydrocarbon Synthesis
28. Catalytic Reforming
29. Alkyl Group
30. Acyclic Hydrocarbon

Chapter 11

Functional Groups

1. Alcohol
2. Ether
3. Aldehyde
4. Ketone
5. Carboxylic Acid
6. Ester
7. Amine
8. Amide
9. Thiol
10. Nitrile
11. Acid Chloride
12. Anhydride
13. Phenol
14. Imine
15. Enamine
16. Sulfide
17. Disulfide
18. Peroxide
19. Phosphate
20. Sulfonic Acid
21. Nitro Group
22. Azo Compound
23. Carbonyl Group
24. Hydroxy Group
25. Alkoxy Group
26. Vinyl Group
27. Allyl Group
28. Phthalimide
29. Halide
30. Aromatic Ring

Chapter 12

Steriochemistry

1. Chiral Center
2. Chirality
3. Enantiomer
4. Diastereomer
5. Racemic Mixture
6. Optical Activity
7. Specific Rotation
8. Stereoisomer
9. Newman Projection
10. Fischer Projection
11. R/S Nomenclature
12. E/Z Nomenclature
13. Stereogenic Center
14. Achiral
15. Conformational Isomer
16. Meso Compound
17. Resolution
18. Optical Isomer
19. Symmetry
20. Asymmetric Synthesis
21. Cahn-Ingold-Prelog Priority Rules
22. Stereoelectronic Effects
23. Pseudo Symmetry
24. Stereospecific Reaction
25. Stereoselective Reaction
26. Conformation
27. Chair Conformation
28. Boat Conformation
29. Axial and Equatorial Positions
30. Ring Flip

Chapter 13

Reaction Mechanisms

1. Nucleophile
2. Electrophile
3. Transition State
4. Intermediate
5. Reaction Coordinate
6. Mechanistic Pathway
7. Nucleophilic Substitution
8. Electrophilic Addition
9. Free Radical Reaction
10. Elimination Reaction
11. Addition Reaction
12. Rearrangement Reaction
13. SN1 Mechanism
14. SN2 Mechanism
15. E1 Mechanism
16. E2 Mechanism
17. Radical Mechanism
18. Concerted Reaction
19. Homolytic Cleavage
20. Heterolytic Cleavage
21. Activation Energy
22. Rate-Determining Step
23. Catalysis
24. Lewis Acid
25. Lewis Base
26. Bronsted Acid
27. Bronsted Base
28. Acid-Base Reaction
29. Solvolysis
30. Hydrolysis
31. Condensation Reaction
32. Diels-Alder Reaction
33. Aldol Condensation

34. Grignard Reaction
35. Wittig Reaction
36. Michael Addition
37. Electrophilic Aromatic Substitution
38. Rearrangement
39. Coupling Reaction
40. Decomposition Reaction

Chapter 14

Additonal Terms

1. Polymerization
2. Copolymer
3. Cross-Linking
4. Hydrolysis
5. Thermodynamics
6. Kinetics
7. Equilibrium
8. Le Chatelier's Principle
9. Reaction Mechanism
10. Reaction Energy Profile
11. Synthesis
12. Isolation
13. Purification
14. Chromatography
15. Spectroscopy
16. Mass Spectrometry
17. Titration
18. Stoichiometry
19. Concentration
20. Dilution
21. pH
22. Buffer
23. Ionization
24. Solubility

25. Solvent
26. Reagent
27. Product
28. Byproduct
29. Yield
30. Selectivity
31. Specificity
32. Functionalization
33. Stereoinversion
34. Reaction Yield
35. Atom Economy
36. Synthesis Route
37. Side Reaction
38. Mechanistic Study
39. Experimental Design
40. Quantum Chemistry

PHYSICAL CHEMISTRY

Chapter 15

Thermodynamics

1. Temperature
2. Heat
3. Work
4. Internal Energy
5. Enthalpy
6. Entropy
7. Gibbs Free Energy
8. Helmholtz Free Energy
9. First Law of Thermodynamics
10. Second Law of Thermodynamics
11. Third Law of Thermodynamics
12. Reversible Process
13. Irreversible Process
14. Closed System

15. Open System
16. Isolated System
17. State Function
18. Path Function
19. Adiabatic Process
20. Isothermal Process
21. Isobaric Process
22. Isochoric Process
23. Phase Equilibrium
24. Chemical Potential
25. Partial Molar Free Energy
26. Clausius-Clapeyron Equation
27. Carnot Cycle
28. Heat Capacity
29. Molar Heat Capacity
30. Standard State
31. Standard Enthalpy of Formation
32. Standard Entropy
33. Phase Diagram
34. Vapor Pressure
35. Critical Point
36. Latent Heat
37. Heat of Reaction
38. Hess's Law
39. Joule-Thomson Effect
40. Enthalpy of Vaporization
41. Enthalpy of Fusion
42. Reaction Coordinate
43. Transition State
44. Activation Energy
45. Spontaneity
46. Thermodynamic Equilibrium

Chapter 16

Chemical Kinetics

1. Reaction rate
2. Rate law
3. Rate constant (k)
4. Order of reaction
5. Zero-order reaction
6. First-order reaction
7. Second-order reaction
8. Half-life
9. Activation energy (Ea)
10. Arrhenius equation
11. Catalysts
12. Reaction mechanism
13. Transition state
14. Elementary reaction
15. Intermediate species
16. Collision theory
17. Molecularity
18. Steady-state approximation
19. Rate-determining step
20. Chain reaction
21. Enzyme kinetics
22. Michaelis-Menten equation
23. Homogeneous catalysis
24. Heterogeneous catalysis
25. Reaction coordinate
26. Energy profile diagram
27. Pre-exponential factor
28. Concentration effect
29. Temperature effect
30. Integrated rate law
31. Complex reactions
32. Parallel reactions
33. Consecutive reactions

Chapter 17

Chemical Equilibrium

15. Catalyst effect
16. Gibbs free energy (ΔG)
17. Exergonic reaction
18. Endergonic reaction
19. Standard equilibrium constant ($K°$)
20. Equilibrium position
21. Solubility product (Ksp)
22. Common ion effect
23. Buffer solutions
24. Ionization
25. Degree of dissociation
26. Phase equilibrium
27. Equilibrium shift
28. Reaction rate equality

Chapter 18

Electrochemistry

1. Redox reactions
2. Oxidation
3. Reduction
4. Oxidizing agent
5. Reducing agent
6. Electrochemical cell
7. Galvanic cell (Voltaic cell)
8. Electrolytic cell
9. Electrode
10. Anode
11. Cathode
12. Salt bridge
13. Cell potential (Electromotive force, EMF)
14. Standard electrode potential ($E°$)
15. Nernst equation
16. Electrolysis
17. Faraday's laws of electrolysis

18. Electroplating
19. Conductivity
20. Molar conductivity
21. Concentration cell
22. Fuel cell
23. Corrosion
24. Half-cell reactions
25. Battery
26. Primary cell
27. Secondary cell (Rechargeable battery)
28. Overpotential
29. Ion-selective electrode

ANALYTICAL CHEMISTRY

Chapter 19

1. Qualitative Analysis
2. Cation analysis
3. Anion analysis
4. Precipitation reaction
5. Solubility product (Ksp)
6. Flame test
7. Confirmation test
8. Spot test
9. Complexation reactions
10. Reagent
11. Separation techniques
12. Filtration
13. Centrifugation
14. Decantation
15. Acid-base reactions
16. Selective precipitation
17. Group reagents
18. Ion identification
19. Qualitative inorganic analysis

20. Chromatography
21. Test for gases (e.g., H_2, O_2, CO_2)

Chapter 20

Quantitative Analysis

1. Titration
2. Standard solution
3. Molarity (M)
4. Normality (N)
5. Primary standard
6. Secondary standard
7. End point
8. Equivalence point
9. Indicator
10. Gravimetric analysis
11. Volumetric analysis
12. Calibration curve
13. Limit of detection (LOD)
14. Limit of quantification (LOQ)
15. Percent purity
16. Analytical balance
17. Spectrophotometry
18. Beer's Law
19. Precision
20. Accuracy

Chapter 21

Spectroscopy

1. Absorbance
2. Transmittance
3. Wavelength
4. Spectrometer
5. Spectrum

6. Molar absorptivity (extinction coefficient)
7. Calibration curve
8. Beer's Law
9. Infrared (IR) spectroscopy
10. Ultraviolet-visible (UV-Vis) spectroscopy
11. Nuclear magnetic resonance (NMR) spectroscopy
12. Mass spectrometry (MS)
13. Raman spectroscopy
14. Chromophore
15. Solvent effects
16. Resolution
17. Quantitative analysis
18. Qualitative analysis
19. Sample preparation
20. Fourier-transform infrared (FTIR) spectroscopy

Chapter 22

Chromatography

1. Stationary phase
2. Mobile phase
3. Retention factor (Rf)
4. Column chromatography
5. Thin-layer chromatography (TLC)
6. Gas chromatography (GC)
7. High-performance liquid chromatography (HPLC)
8. Supercritical fluid chromatography (SFC)
9. Chromatographic resolution
10. Partition coefficient
11. Gradient elution
12. Sample introduction
13. Detector
14. Baseline
15. Elution
16. Separation

BIOCHEMISTRY

Chapter 23

Proteins

Chapter 24

Carbohydrates

1. Monosaccharides
2. Disaccharides
3. Oligosaccharides
4. Polysaccharides
5. Glycosidic Bonds
6. Starch
7. Glycogen
8. Cellulose
9. Chitin
10. Carbohydrate Metabolism
11. Glycolysis
12. Gluconeogenesis
13. Pentose Phosphate Pathway
14. Insulin
15. Glucagon
16. Blood Sugar Levels
17. Hexokinase
18. Fructose
19. Galactose
20. Sorbitol
21. Mannose
22. Fucose
23. Glycoconjugates
24. Lectins
25. Glycobiology

Chapter 25

Lipids

Chapter 26

Nucleic Acids

4. Adenine
5. Thymine
6. Cytosine
7. Guanine
8. Uracil
9. Replication
10. Transcription
11. Translation
12. mRNA (Messenger RNA)
13. tRNA (Transfer RNA)
14. rRNA (Ribosomal RNA)
15. Codon
16. Anticodon
17. Chromosomes
18. Histones
19. Gene
20. Alleles
21. Mutation
22. Intron
23. Exon
24. Gene Expression
25. CRISPR

ADVANCED TOPICS

Chapter 27

Quantum Chemistry

1. Electron
2. Energy quantization
3. Electron spin
4. Pauli exclusion principle
5. Born interpretation
6. Heisenberg's uncertainty principle
7. Atomic orbitals

8. Double-slit experiment
9. Wave function
10. Quantum tunneling
11. Spin-statistics theorem
12. Orbital diagram
13. Quantum superposition
14. Quantum entanglement
15. Building-up principle (Aufbau principle)
16. Resonance theory
17. Frontier molecular orbital theory
18. Spin state
19. Quantum harmonic oscillator
20. Density Functional Theory (DFT)

Chapter 28

Surface Chemistry

1. Van der Waals forces
2. Surfactant
3. Chemisorbed layer
4. Adsorption
5. Triple point
6. Contact angle
7. Langmuir film
8. Capillary action
9. Two-dimensional gas
10. Spreading
11. Surface tension
12. Adsorption isotherm
13. Helmholtz layer
14. Desorption
15. Hydrophobicity
16. Soret effect
17. Langmuir adsorption model
18. Photoelectric effect

19. Epitaxial growth
20. Coefficient of drag

Chapter 29

Ploymer Chemistry

1. Polymer
2. Polymerization-depolymerization equilibrium
3. Step-growth polymerization
4. Branched polymer
5. Degree of polymerization
6. Copolymer
7. Glass transition temperature
8. Ziegler-Natta catalyst
9. Lower critical solution temperature (LCST)
10. Free radical polymerization
11. Crosslinking
12. Isotactic polymer
13. Initiator
14. Polymer blend
15. Chain transfer reaction
16. End group
17. Oxidative degradation
18. Living polymerization
19. Molecular weight distribution
20. Curing
21.

Chapter 30

Green Chemistry

1. Prevention of waste (1st principle of Green Chemistry)
2. Atom economy
3. Supercritical CO_2
4. Biocatalysis
5. Photochemistry

6. E-factor (Environmental factor)
7. Room temperature reaction
8. Biomass
9. Lifecycle assessment
10. Heterogeneous nanocatalyst
11. Ionic liquid
12. Microwave-assisted synthesis
13. Biodegradability
14. Bio-based feedstock
15. Designing safer chemicals
16. Solvent-free reaction
17. Natural product chemistry
18. In-situ process monitoring
19. Green Star System
20. Photocatalysis